Introduction to Quantum Information Science

Dedicated to Michael and Mia

Two Riders on The Storm: May you fly without wings and conquer without a sword.

Introduction to Quantum Information Science

Vlatko Vedral

School of Physics and Astronomy
University of Leeds

OXFORD
UNIVERSITY PRESS

Great Clarendon Street, Oxford OX2 6DP

Oxford University Press is a department of the University of Oxford.
It furthers the University's objective of excellence in research, scholarship,
and education by publishing worldwide in

Oxford New York

Auckland Cape Town Dar es Salaam Hong Kong Karachi
Kuala Lumpur Madrid Melbourne Mexico City Nairobi
New Delhi Shanghai Taipei Toronto

With offices in

Argentina Austria Brazil Chile Czech Republic France Greece
Guatemala Hungary Italy Japan Poland Portugal Singapore
South Korea Switzerland Thailand Turkey Ukraine Vietnam

Oxford is a registered trade mark of Oxford University Press
in the UK and in certain other countries

Published in the United States
by Oxford University Press Inc., New York

© Vlatko Vedral, 2006

The moral rights of the author have been asserted
Database right Oxford University Press (maker)

First published 2006

All rights reserved. No part of this publication may be reproduced,
stored in a retrieval system, or transmitted, in any form or by any means,
without the prior permission in writing of Oxford University Press,
or as expressly permitted by law, or under terms agreed with the appropriate
reprographics rights organization. Enquiries concerning reproduction
outside the scope of the above should be sent to the Rights Department,
Oxford University Press, at the address above

You must not circulate this book in any other binding or cover
and you must impose the same condition on any acquirer

British Library Cataloguing in Publication Data

Data available

Library of Congress Cataloging in Publication Data

Data available

Typeset by Newgen Imaging Systems (P) Ltd., Chennai India
Printed in Great Britain
on acid-free paper by
Biddles Ltd., www.biddles.co.uk

ISBN 0–19–9215707   978–0–19–9215706 (Hbk)

1 3 5 7 9 10 8 6 4 2

Preface

This book is a result of a number of courses I have given on the topic of quantum information. In chronological order, they were delivered in Heidelberg, Germany, Merida, Mexico and Vienna, Austria. The book comprises between 22 and 26 lectures (depending on the speed of delivery) and is ideal for an initial study of the subject of quantum information.

I have tried to cover all the topics I believe relevant, but this is of course significantly dependent on my own bias. The level is suitable for third/fourth year undergraduates and I would expect my postgraduate students to be familiar with it at the end of their first year. After introductions to classical information theory and the basics of quantum mechanics (I assume familiarity with neither), four major topics are covered. The first is the theory of quantum communication, the second is the theory of entanglement, the third is quantum algorithms, and the last is error correction. A major topic that is left out is a discussion of the practical implementation of quantum computation. This is covered in far more detail elsewhere and I did not find it suitable for this book.

Quantum information is a fascinating topic precisely because it shows that the laws of information processing are actually dependent on the laws of physics. However, it is also very interesting to see that information theory has something to teach us about physics. Both of these directions will be discussed throughout the book. I hope that readers will find the book interesting. I also hope that, if readers choose to undertake research in quantum information, this book will provide an excellent basis for independent work.

Acknowledgments

This book has grown out of many lectures, and out of interactions with students and researchers in the field of quantum information. First and foremost I am grateful to Artur Ekert, who introduced me to the field of quantum information in 1994/95. I have found it fascinating ever since. I have also greatly benefited from numerous interactions with many people in the field. In is impossible to list all of them here. The ones whose influence can be seen most clearly in this book are Sougato Bose, Časlav Brukner, Artur Ekert, Peter Knight, Goran Lindblad, Mio Murao, Masanao Ozawa, Martin Plenio, and Anton Zeilinger.

Special thanks go to Steve Evans, the Head of Physics and Astronomy at Leeds, who has made my transfer to Leeds very smooth and enjoyable and allowed me the time to work on this project (and many others). Caroline Rogers has (yet again) proven invaluable in helping me compile the book. Her constant input throughout the writing and especially her final touches were crucial for the completion of this project. Cristhian Avila is gratefully acknowledged for his help in finalising the book. I am also very grateful to Salvador Venegas for typing up a very early draft of the notes and drawing some of the diagrams. Luke Rallan and Julian Hartley have also been involved in drawing some diagrams and reading some of the early chapters of the book. The picture on the front cover is by courtesy of NASA/JPL and Caltech.

My family has, as always, provided a constant source of joy. Without Ivona, Mikey, and Mia none of this would make any sense.

Contents

PART I   CLASSICAL AND QUANTUM INFORMATION

1   Classical information   3
    1.1   Information and physics   3
    1.2   Quantifying information   4
    1.3   Data compression   7
    1.4   Related measures of information   8
        1.4.1   Relative entropy   9
        1.4.2   Joint entropy   10
        1.4.3   Conditional entropy   10
        1.4.4   Mutual information   10
    1.5   Capacity of a noisy channel   11
    1.6   Summary   12

2   Quantum mechanics   14
    2.1   Dirac notation   14
    2.2   The qubit, higher dimensions, and the inner product   16
    2.3   Hilbert spaces   17
    2.4   Projective measurements and operations   19
    2.5   Unitary operations   20
    2.6   Eigenvectors and eigenvalues   21
    2.7   Spectral decomposition   22
    2.8   Applications of the spectral theorem   23
    2.9   Dirac notation shorthands   24
    2.10   The Mach–Zehnder interferometer   25
    2.11   The postulates of quantum mechanics   27
    2.12   Mixed states   28
    2.13   Entanglement   29
    2.14   Summary   30

3   Quantum information—the basics   31
    3.1   No cloning of quantum bits   31
    3.2   Quantum cryptography   33
    3.3   The trace and partial-trace operations   35
    3.4   Hilbert space extension   37
    3.5   The Schmidt decomposition   38

	3.6 Generalized measurements	40
	3.7 CP-maps and positive operator-valued measurements	41
	3.8 The postulates of quantum mechanics revisited	42
	3.9 Summary	42
4	**Quantum communication with entanglement**	**44**
	4.1 Pure state entanglement and Pauli matrices	44
	4.2 Dense coding	45
	4.3 Teleportation	46
	4.4 Entanglement swapping	48
	4.5 No instantaneous transfer of information	49
	4.6 The extended–Hilbert–space view	50
	4.7 Summary	50
5	**Quantum information I**	**52**
	5.1 Fidelity	53
	5.2 Helstrom's discrimination	54
	5.3 Quantum data compression	55
	5.4 Entropy of observation	58
	5.5 Conditional entropy and mutual information	59
	5.6 Relative entropy	61
	5.7 Statistical interpretation of relative entropy	62
	5.8 Summary	66
6	**Quantum information II**	**68**
	6.1 Equalities and inequalities related to entropy	68
	6.2 The Holevo bound	71
	6.3 Capacity of a bosonic channel	73
	6.4 Information gained through measurements	75
	6.5 Relative entropy and thermodynamics	76
	6.6 Entropy increase due to erasure	77
	6.7 Landauer's erasure and data compression	78
	6.8 Summary	78

PART II   QUANTUM ENTANGLEMENT

7	**Quantum entanglement—introduction**	**81**
	7.1 The historical background of entanglement	81
	7.2 Bell's inequalities	83
	7.3 Separable states	85
	7.4 Pure states and Bell's inequalities	86
	7.5 Mixed states and Bell's inequalities	87
	7.6 Entanglement in second quantization	87
	7.7 Summary	91
8	**Witnessing quantum entanglement**	**92**
	8.1 Entanglement witnesses	93
	8.2 The Jamiolkowski isomorphism	95

	8.3	The Peres–Horodecki criterion	97

	8.3	The Peres–Horodecki criterion	97
	8.4	More examples of entanglement witnesses	99
	8.5	Summary	100
9	**Quantum entanglement in practice**		102
	9.1	Measurements with a Mach–Zehnder interferometer	102
	9.2	Interferometric implementation of Peres–Horodecki criterion	104
		9.2.1 Measuring tr ϱ^2?	104
		9.2.2 Generalization to tr ϱ^k	105
		9.2.3 Measuring tr $(\varrho^{T_2})^k$	106
	9.3	Measuring the fidelity between ϱ and σ	106
	9.4	Summary	107
10	**Measures of entanglement**		108
	10.1	Distillation of multiple copies of a pure state	108
	10.2	Analogy with the Carnot Cycle	110
	10.3	Properties of entanglement measures	111
	10.4	Entanglement of pure states	113
	10.5	Entanglement of mixed states	113
	10.6	Measures of entanglement derived from relative entropy	117
	10.7	Classical information and entanglement	121
	10.8	Entanglement and thermodynamics	123
	10.9	Summary	128

PART III   QUANTUM COMPUTATION

11	**Quantum algorithms**		131
	11.1	Computational complexity	131
	11.2	Deutsch's algorithm	133
		11.2.1 Deutsch's algorithm and the Holevo bound	135
	11.3	Oracles	136
	11.4	Grover's search algorithm	137
	11.5	Quantum factorization	140
		11.5.1 Factorization	141
		11.5.2 The quantum Fourier transform	142
		11.5.3 Phase estimation	144
	11.6	Summary	145
12	**Entanglement, computation and quantum measurements**		146
	12.1	Optimization of searches using entanglement	147
	12.2	Model for quantum measurement	149
	12.3	Correlations and quantum measurement	151
	12.4	The ultimate limits of computation: the Bekenstein bound	157
	12.5	Summary	158
13	**Quantum error correction**		160
	13.1	Introduction	160
	13.2	A simple example	160

13.3	General conditions	162
13.4	Reliable quantum computation	165
13.5	Quantum error correction considered as a Maxwell's demon	167
	13.5.1 Pure states	171
	13.5.2 Mixed states	172
13.6	Summary	173

14 Outlook   175

Bibliography   179

Index   181

Part I

Classical and Quantum Information

1
Classical information

This book deals with the basic concept of information, its importance, and the similarities and differences between classical and quantum information. The contents and scope can be summarized as follows. First we discuss classical information theory and its application to communication. We then discuss quantum mechanics and communication based on the laws of quantum mechanics. Entanglement is a key resource in quantum communication. We go on to discuss entanglement both from the fundamental perspective and from the perspective of its information-processing capability. Finally, we introduce quantum computation (in the context of the search and factorization algorithms) and the basics of quantum error correction. Throughout this book, the connections between information theory, thermodynamics, and (quantum) physics are emphasized and discussed.

A good reference book (but far too large to ever read) is Quantum Computation and Quantum Information (Nielsen and Chuang 2002). There are also many lecture notes on the Internet which can be downloaded (for free), including John Preskill's physics notes at http://www.theory.caltech.edu/people/preskill/ph229/, the Carnegie Mellon course at http://quantum.phys.cmu.edu/QCQI/index1.html, which has links to other courses, and the Cambridge University course at http://cam.qubit.org/lectures/qitheory.php. There are also various survey articles available on quantum information, for example, my review of a few years back (Vedral 2002).

1.1 Information and physics

During the second half of the twentieth century, the development of computer science led to a new way of understanding physics—in very general terms, physical systems can be thought of as computers processing information. The initial state of a physical system is the input to a computer and evolves as the computer performs computations to some final state, which is the computer's output. Studying physics in this way not only allows us to use the tools of theoretical computer science and information theory to understand physical laws, but also provides us with an entirely new way of thinking about physics—we can think of the values of the physical attributes of a system as information held in that system. On the other hand, the fact that physics and computation can in some sense be thought of as synonymous means that the laws of information processing depend fully on the laws of physics. This means that whenever we discover new laws of physics,[1] the laws of computation will most likely also be affected in a very profound way.

[1] This hasn't happened for about 80 years now.

4 *Classical information*

Classical information theory assumes that information evolves according to the laws of classical (Newtonian) physics (the pioneers of information theory did not even make this assumption explicitly —it was just there naturally, because classical physics is very intuitive). We begin in this chapter by studying the basics of classical information theory—how information can be quantified, how data compression works, and how we can compare the information in two sets of data. This provides the groundwork for the later chapters, where we study quantum information theory, in which information is represented by quantum states and allowed to evolve according to the laws of quantum mechanics.

I should say at the outset that I regard physics as the queen and mathematics as one of its (many) servants—other servants being subjects such as philosophy and logic[2]. This book, if anything, is a testimony to the fact that physics is more fundamental, basic and important than mathematics. Initially it was thought (by the ancient Greeks, for instance) that geometry is part of mathematics until Einstein showed that geometry is determined by gravity in the universe. So, in order to determine what kind of geometry we live in, we need to measure angles and distances in our vicinity (i.e. We need to study the physics of General Relativity). Now, a century after Einstein's great insight, we understand that logic, information, and computation are not parts of mathematics either, but also belong to physics. How much information we can store and how quickly we can process it depends (only) upon the laws of physics (and not on some arbitrary axioms of mathematics that come out of nowhere). This is exactly what this book is all about.

1.2 Quantifying information

How can information be quantified? One approach is to agree on a context in which the information is to be used and then quantify how useful the information is within that context.[3] The context in which the information is to be used defines the properties we would like the information measure to possess.

Shannon developed the first successful theory of information in 1948 in the context of sending information over a channel such as a telephone wire. Although his main motivation was to maximize telephone communication profits for the Bell Corporation, namely he wanted to investigate the maximum communication capacity of a (classical) communication channel, his theory is so general that it can be (and has been) applied to many diverse disciplines. It is used in biology (especially genetics), sociology and economics, to name a few.[4] It is also used in physics, and this usage will be the main topic of this book. We shall look at information from two different perspectives: we shall investigate what the laws of physics have to say about information processing and what the laws of information processing have to say about physics (if anything, of course).

---

[2]When I lectured at the Schrödinger institute in Vienna, this view—which I stated clearly at the outset—didn't go down well with some people in the audience (mathematicians, I guess). Most of them, nevertheless, stayed on to hear the rest of the course, which I believe was quite popular all in all.

[3]This is the physical situation where information is conveyed and will be discussed shortly in great detail.

[4]It has also been misused even more than this...

Shannon modeled information as events which occur with certain probabilities.[5] He postulated the following requirements that any measure of information must possess:

- *The amount of information in an event x must depend only upon its probability p* This is a very natural requirement, since the more surprised we are by an event happening, the more information this event carries.[6] For example, a fall of snow in the Arctic carries very little information, whereas snow in the Amazon rain forest could be a very significant event.
- *$I(p)$ is a continuous function of p* Continuity is a desirable quality for physical quantities (disregarding phase transitions and some other anomalies). In this case, the continuity condition says that if the probability of an event changes by a very small amount, the information contained in that event changes by only a small amount. Given the previous postulate, it makes little sense that a small change in the probability of some event would lead to us being much more surprised when the event takes place.
- *$I(p_x, p_y) = I(p_x) + I(p_y)$* This additivity assumption is the most stringent one, in the sense that not many functions of real positive numbers (i.e. probabilities) are additive. Why do we want additivity? If we have two independent events where the first one happens with probability p_1 and the second with probability p_2, then additivity says that the information carried in both events together, $\mathbf{I}(p_x, p_y)$, is the sum of the information carried in each of the two events, $\mathbf{I}(p_x) + \mathbf{I}(p_y)$. I would like to warn the reader that this assumption can be given up, but then the information measure will also change. Such measures exist; however Shannon's information measure is much more popular and far more widely used.[7]

Shannon proved that there is a unique measure which satisfies these requirements. This measure is unique up to an additive and multiplicative constant (i.e. an affine transformation). The measure is called the Shannon entropy and is defined as follows. If X is a random variable over a set of events x such that event x occurs with probability p_x, then the Shannon entropy of event x is $-\log_2(p_x)$.

The proof of the uniqueness of this measure proceeds as follows. We use the third property to show that $I(p_x^r) = rI(p_x)$ for any rational r, and then express a probability p_x as a power of one-half, $p_x = 2^{-r}$. We use the second property to show that the proof holds not only for rational but also for real r. The first property is used in the proof to talk about the information in events solely in terms of their probabilities.

We now present the proof in more detail. Let $I(p_x^s)$ stand for the information contained in s independent occurrences of an event x. The additivity property shows that the information contained in s independent occurrences of x is s times the information that occurs in a single event x:

---

[5]This is very different from what we think information is in ordinary everyday usage. In everyday information, probabilities may not matter as much as the meaning of the words and gestures and the person communicating with us. All this is completely irrelevant when it comes to Shannon's definition.

[6]This was already realized by the ancient Greeks. The significant thing is that nothing else should enter a quantity measuring information.

[7]This is because Shannon's measure has a strong physical motivation and meaning; other measures are more or less useless, with a few notable exceptions.

$$
\begin{align}
I(p_x^s) &= I(p^{s-1}, p_x) \tag{1.1}\\
&= I(p_x^{s-1}) + I(p_x) \tag{1.2}\\
&= I(p_x^{s-2}, p_x) + I(p_x) \tag{1.3}\\
&= I(p_x^{s-2}) + I(p_x) + I(p_x) \tag{1.4}\\
&= \ldots \tag{1.5}\\
&= sI(p_x) \,. \tag{1.6}
\end{align}
$$

To show that $I(p_x^r) = rI(p_x)$ holds for any rational $r = s/t$ where s and t are integers, we can use eqn 1.6 to show that $I(p_x^{1/t}) = (1/t)I(p_x)$:

$$
\begin{align}
I(p_x^{1/t}) &= \frac{1}{t} \times (t \times I(p_x^{1/t})) \tag{1.7}\\
&= \frac{1}{t} \times I(p_x^{t \times (1/t)}) \tag{1.8}\\
&= \frac{1}{t} \times I(p_x) \tag{1.9}
\end{align}
$$

Any rational r can be written as $r = s/t$ where s and t are integers. Equations 1.6 and 1.9 together show that for any rational r,

$$
\begin{align}
I(p_x^r) &= I(p_x^{s/t}) \tag{1.10}\\
&= \frac{s}{t} \times I(p_x^{s/t}) \tag{1.11}\\
&= rI(p_x) \,. \tag{1.12}
\end{align}
$$

We now use continuity to complete the proof. Any probability p_x of an event can be expressed as $p_x = 2^{-\log(p_x)}$, where $\log(p_x)$ is arbitrarily close to a rational number r. Since I is continuous, we have

$$
\begin{align}
I(p_x) &= I(2^{-\log(p_x)}) \tag{1.13}\\
&= -\log(p_x) I\left(\frac{1}{2}\right) \,. \tag{1.14}
\end{align}
$$

Defining $I(1/2) = 1$ (we stated above that the Shannon entropy is unique only up to a constant multiplicative or additive factor), we find that $I(p_x) = -\log_2(p_x)$ as required.

The Shannon entropy is of central importance in classical information theory (and also in classical statistical mechanics via Boltzmann's statistical formula for entropy—this is in fact why Shannon called his quantity "entropy").[8] Interestingly, Shannon's logic in using entropy because of its additive property is reminiscent of Maxwell's argument in deriving his velocity distribution for a gas of atoms (or molecules). Briefly, he said that the velocity probability density must be a function of energy, i.e. $f(v^2)$,

---

[8] The real story behind this is that von Neumann advised Shannon to use the name "entropy" since no one knows what entropy is, so Shannon would always have an edge in any discussion. Note that the term "information" would not achieve anywhere near the same effect.

where v is the velocity. However, the probability distributions in the x, y, and z directions must be independent and so we must have that

$$f(v_x^2 + v_y^2 + v_z^2) = f(v_x^2)f(v_y^2)f(v_z^2) \,. \tag{1.15}$$

The only function that satisfies the above equation is the exponential and so $f(v) = Ae^{Bv^2}$, i.e. the Maxwell–Boltzmann distribution! You see how far simple logic can take us in general ...[9]

1.3 Data compression

Data compression is an example of an application of the Shannon entropy. It will now become clearer why the Shannon entropy is useful in the theory of communication. We model a set of data to be compressed by a random variable X. The Shannon entropy of X is then defined as the expected logarithm of the probability:

$$H(X) = -\sum_x p(X = x) \log(P(X = x)) \,. \tag{1.16}$$

It is this form of Shannon entropy that we use for data compression.

Suppose Alice tosses a biased coin n times and wants to send Bob the result of her coin tosses. From previous experiments (of which they have performed zillions and zillions), Alice and Bob have found that each coin toss produces a head with probability 0.75 and a tail with probability 0.25. They say that the random variable X represents the output of each coin toss, where $P(X = heads) = 0.75$ and $P(X = tails) = 0.25$. If Alice now tosses the coin a large number of times, say 1 million, Alice and Bob can be sure that the coin will output about 750 000 heads and 250 000 tails—this is called a *typical sequence*. As n gets bigger, they can be even more sure of getting a typical sequence containing about $0.75n$ heads and $0.25n$ tails. We shall show later on that the number of such typical sequences is approximately $2^{nH(X)}$. If Alice wants to send Bob the results of her coin tosses, she can just tell him which of the typical sequences she has by sending him a string x. Each bit of the string x has two values, so a string of length m has 2^m possible values. There are about $2^{nH(X)}$ typical sequences so Alice can use a string of length approximately $nH(X)$ to tell Bob which of these typical sequences she has. As n increases towards infinity, the probability of getting a typical sequence increases towards 1. With a small probability, the coin will produce an atypical sequence (i.e. a sequence which is not typical), in which case the compression fails. For this reason, this compression is often called *lossy compression*.

In general, the aim of data compression is for two parties, which we call Alice and Bob, to communicate over a (noiseless) channel. Alice has a set of data represented by a probability distribution X, from which she wants to send a large block of n messages to her friend Bob using as few bits as possible. How data compression works is that a large block of information can be divided up into typical sequences containing many probable characters, and atypical sequences containing long sequences of improbable characters. By encoding only the typical sequences, far fewer bits can be used to encode

---

[9] I should say, more precisely, that this is how far simple symmetry considerations can take us; "logic", of course, has nothing to do with it.

8 *Classical information*

the entire message. With a very small probability, the random variable may generate an atypical sequence which has not been encoded and the compression fails. For this reason, this compression is often called **lossy compression**.

We now go over the proof the optimal rate of lossy compression is the Shannon entropy of the data to be compressed more formally. Let X^n represent n independent samples from a random variable X and let $x_1 \ldots x_n$ be a sequence of values obtained from X^n. $H(X)$ can be formulated as the expectation value of $-\log(p(x))$, which we write as $-E(\log(p(x)))$. The weak law of large numbers tells us that for any $\delta > 0$ and $\epsilon > 0$, there exists a sufficiently large n such that, with a probability of at least $1 - \epsilon$, an ϵ-typical sequence $x_1 \ldots x_n$ occurs satisfying

$$2^{-n(H(X)-\delta)} < p(x_1, \ldots, x_n) < 2^{-n(H(X)+\delta)} . \tag{1.17}$$

Using the bound above and the fact that the total probability of all ϵ-typical sequences is between $1-\epsilon$ and 1, we find the following bound on the number of sequences $N(\epsilon, \delta)$:

$$2^{-H(X)+\delta} \geq N(\epsilon, \delta) \geq (1-\epsilon)2^{-H(X)-\delta} . \tag{1.18}$$

Using $n(H(X) + \delta)$ bits, all the ϵ-typical sequences can be encoded with an error of at most ϵ; ϵ and δ can be chosen to be arbitrarily small by increasing n, so that the rate of compression can be arbitrarily close to $H(X)$ bits and with as small an error ϵ as desired, which is why we say that the rate of compression is equal to the Shannon entropy of X.

What happens if we try to encode X using fewer bits? If $R < (1-\epsilon)2^{-H(X)-\delta}$ bits are used, then as n grows large, 2^{nR} becomes significantly smaller than $2^{-nH(X)}$, so most of the typical sequences are not encoded and the probability of error increases. Note that here, unlike in the case of human speech for example, it is the most random sequence that is least compressible (in fact, it is not compressible at all, since it has the maximum entropy).[10]

1.4 Related measures of information

The Shannon entropy can be used to define other measures of information which capture the relationships between two random variables X and Y. Four such measures are the following:

- the *relative entropy*, which measures the similarity between two random variables;
- the *joint entropy*, which measures the combined information in two random variables;
- the *conditional entropy*, which measures the information contained in one random variable given that the outcome of another random variable is known;
- the *mutual information*, which measures the correlation between two random variables, in terms of how much knowledge of one random variable reduces the information gained from learning the other random variable.

---

[10]Pages and pages of convoluted text in a randomly chosen political manifesto can frequently be compressed to a single sequence, e.g. "we shall have to increase taxes". But then, unlike Shannon information, manifestos are meant to increase misinformation, rather than channel capacity.

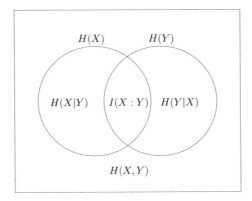

Fig. 1.1 The relationship between the joint entropy $H(X,Y)$, the conditional entropy $H(X|Y)$ and the mutual information $I(X:Y)$. From this diagram, we can see how the different entropies are related by addition; for example, $H(X,Y) = H(X|Y) + I(X:Y) + H(Y|X)$.

These measures are very useful tools for thinking about information and are interrelated by many equalities. We shall informally describe the relationships and leave the formal proofs as an exercise. The relationships between the joint, conditional, and mutual entropies are shown in Fig. 1.1.

1.4.1 Relative entropy

The relative entropy is defined as follows:

$$H(X||Y) = -\sum_{x,y} p(x)\log(p(y)) - H(X) \quad (1.19)$$

$$= \sum_{x,y} p(x)\log p(x)/p(y) . \quad (1.20)$$

The relative entropy represents the difference between the expected information obtained from the events of Y given that they are distributed according to X (i.e. $\sum_{x,y} p(x)\log(p(y))$) and the expected information obtained from X (i.e. $H(X)$). The relative entropy is always positive. If $X = Y$ then the relative entropy between X and Y is zero. The relative entropy increases as the distance between X and Y increases.

The relative entropy was of great importance in the classical statistical mechanics of Gibbs and is very useful in quantum information theory, as many key results follow from its monotonic properties (to be discussed later on). The relative entropy expresses the entropic difference between two random variables. This is, in fact, the most proper way of talking about information which is always a relative concept—the uncertainty in a variable is always measured with respect to another variable. The Shannon entropy is a special case of the relative entropy. The Shannon entropy of a random variable is the entropy relative to a state which is known with certainty, i.e. $H(X) = H(X||Y)$, where $P(Y = y) = 1$ for some value y.

1.4.2 Joint entropy

The joint entropy of two random variables X and Y is simply the entropy of the joint distribution of X and Y:

$$H(X,Y) = -\sum_{x,y} p(x,y) \log(p(x,y)) \qquad (1.21)$$

. If X and Y are independent, then the additive property of the Shannon entropy shows that

$$H(X,Y) = H(X) + H(Y) \qquad (1.22)$$

. This equality does not hold when X and Y are not independent.

1.4.3 Conditional entropy

The conditional entropy measures the information gained from learning the outcome of X given that Y is known:

$$H(X|Y) = -\sum_{x,y} p(x|y) \log(p(x|y)) \,, \qquad (1.23)$$

where $p(x|y) = p(x,y)/p(x)$. I learn nothing from looking at the color of the sock on my right foot if I know the color of the sock on my left foot, simply because they are always the same color and so the conditional entropy between the colors of my socks is 0. On the other hand, if X and Y are completely independent then $H(X|Y) = H(X)$ since the information gained from X is not reduced by knowledge about Y.

I can describe the joint distribution of X and Y by describing X and then describing Y given that X is already known. This leads to a relationship between the conditional and joint entropies,

$$H(X,Y) = H(X) + H(Y|X) \,, \qquad (1.24)$$

which can be proved by applying the definition of $p(x|y)$.

1.4.4 Mutual information

The mutual information between two random variables X and Y is the difference between the information gained from learning X and the information gained from learning X when Y is already known.

$$I(X:Y) = H(X) - H(X|Y) \,. \qquad (1.25)$$

If X and Y are independent, then the mutual information between X and Y is zero. If they are completely correlated (such as the colors of my socks), then the mutual information between them is the same as the information contained in X. More formally, if there exists a bijection f such that $P(X = x) = P(Y = f(x))$, then $I(X:Y) = H(X) = H(Y)$.

The mutual information is symmetric—the mutual information between X and Y is the same as the mutual information between Y and X:

$$I(X:Y) = H(Y) - H(Y|X) = I(X:Y) \,. \qquad (1.26)$$

The mutual information between X and Y is the difference between the number of bits required to express X and Y separately and as a joint distribution:

$$I(X:Y) = H(X) + H(Y) - H(X,Y) \,. \tag{1.27}$$

It will be very useful for us later in the book to know that the mutual information between X and Y is the same as the relative entropy between the joint distribution XY and the two distributions X and Y. How correlated X and Y are is measured by how far away the joint distribution XY is from the product of the "marginals" X and Y.

1.5 Capacity of a noisy channel

When Alice talks to Bob, the signal that is sent (the string of bits) can be thought of as a random variable, say X. X is given fully by the probability that Alice sends Bob the string x. If the channel is noisy, this string will in general be changed (owing to errors, which will occasionally toss 0's into 1's and vice versa), as in Fig. 1.2. The changed string, when received by Bob, will therefore be another random variable, say Y. How correlated X and Y are is in fact the same as the channel capacity— the expected number of bits about X that Bob can learn.[11] If, when Alice sends one bit, Bob receives exactly this bit (i.e. We have a noiseless channel), then the correlation between them is also one bit and so is the capacity. If this bit has a chance of being tossed with probability one-half, then X and Y are completely uncorrelated (we have a randomizing channel) and the correlation is zero and so is the capacity for communication— Bob cannot guess anything about Alice's original message. These are two extremes that follow from the following completely general result:

Shannon's theorem. If R is the rate of information production, then, providing that $R < C = I(X:Y)$, the information can be transmitted with an arbitrary reliability.[12]

A source X with entropy H will generate about $2^{nH(X)}$ typical sequences in n steps ("about" indicates that this is only true to within an arbitrary factor as n becomes large). Now, each of these sequences will be read by Bob and each output could then be produced by about $2^{nH(X|Y)}$ inputs, since $H(X|Y)$ represents the entropy of X once Y has been measured. Therefore the total number of useful (nonredundant) messages (as they are called in the field of communications) is

$$N = 2^{n(H(X) - H(X|Y))} \,, \tag{1.28}$$

and therefore, for the capacity, we choose a source with an entropy that maximizes $H(X) - H(X|Y)$, and this is just the mutual information $I(X:Y)$. If instead we choose

---

[11]This scenario is extremely general. Alice could, for example, be the words in this book, Bob could be reader's brain, and the noisy channel could be typographical errors in the book. It is this generality that has led to many beautiful applications, but also to many unfortunate misplaced instances.

[12]Here we present only an intuitive argument to justify the above form of the capacity. We stress that this proof is valid only for ergodic, stationary sources for which most sequences of n bits from a source with an entropy H have a probability of about e^{-nH}. Loosely speaking, a source is stationary if the probabilities of emitting the various possible states do not change over time; it is ergodic if each subsequence of states appears in longer sequences with a frequency equal to its probability. This statement is then an information-theoretic analogue of the Law of Large Numbers in probability theory.

12 Classical information

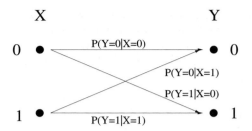

Fig. 1.2 Alice tries to send X to Bob; however, the channel through which she tries to send X causes some errors or noise. The distribution Bob actually receives is Y.

a source whose entropy produces a larger quantity than the channel capacity, then the channel will not be able to handle the input, and errors will inevitably result. The mutual information between the input and the output of a communication channel is thus a very important quantity since it determines the maximum rate at which information can be transmitted without errors occurring.

1.6 Summary

Shannon proposed three requirements that any measure of information I should possess:

1. The amount of information I in an event x must depend only on its probability p.
2. I is a continuous function of the probability.
3. I is additive.

These requirements lead to a unique measure of information (up to an affine transformation, i.e. a multiplicative/additive constant factor), called the Shannon entropy. The Shannon entropy of an event which occurs with probability p is $-\log(p)$. The Shannon entropy of a random variable X is

$$H(X) = -\sum p(X = x) \log(p(X = x)) . \tag{1.29}$$

Data distributed according to X^n (i.e. n copies of X) can be transmitted with an arbitrarily small probability of error as n becomes large by sending $nH(X)$ bits. This is called noiseless data compression.

Some related measures of information are the relative entropy $H(.||.)$, the joint entropy $H(.,.)$, the conditional entropy $H(.|.)$ and the mutual information $I(. : .)$, defined as follows:

$$H(X||Y) = \sum_{x,y} p(x)(\log(p(x)) - \log(p(y))) , \tag{1.30}$$

$$H(X,Y) = -\sum_{x,y} p(x,y) \log(p(x,y)) , \tag{1.31}$$

$$H(X|Y) = -\sum_{x,y} p(x|y) \log(p(x|y)) , \tag{1.32}$$

$$I(X : Y) = H(X) - H(X|Y) . \tag{1.33}$$

The relationships between these measures are shown in Fig. 1.1.

n copies of a random variable X over a noisy channel, that transforms the distribution of X into a distribution Y, can be transmitted with an arbitrarily small probability of error as n becomes large by sending $nC = nI(X:Y)$ bits. C is called the channel capacity.

2
Quantum mechanics

This chapter reviews all the rules of quantum mechanics and their mathematical notation. We also begin to talk about quantum entanglement, which is a fundamental resource in quantum information. I'll do my best to make the treatment independent of any interpretational issues (which, of course, are also exciting and important, but I don't want to "poison" the reader at this early stage![1]).

There are four basic postulates of quantum mechanics which tell you how to represent physical systems, how to represent observations, how to carry out measurements and how systems evolve when "not measured". None of these rules are obvious or natural[2]. They cannot be inferred from anything deeper, as far as we know (although there have been many worthwhile attempts[3]). Before describing the laws of quantum mechanics, we shall explain their mathematical background and describe a Mach–Zehnder interferometer experiment, in which the (strange) readings on the detectors can be explained by quantum mechanics.

2.1 Dirac notation

Dirac notation is a convenient tool used for representing vectors in quantum mechanics. States of physical systems are represented by vectors in quantum mechanics, which is why we need to talk a bit about linear algebra. Why this is so will become apparent later when we discuss some experiments.

Dirac notation is made up of *bras* $\langle .|$ and *kets* $|.\rangle$ which together form *brakets*[4] $\langle .||.\rangle$. Kets can be represented by column vectors, which are familiar to most people. In terms of Cartesian x-y coordinates, a point $p = (a, b)$ can be represented as a column vector

$$\mathbf{p} = \begin{pmatrix} a \\ b \end{pmatrix} \qquad (2.1)$$

---

[1] I shall, of course, gladly indoctrinate the reader towards the end of the book, but by that time they will hopefully have developed much resistance to scientific propaganda.

[2] In fact, even the number "four" is not universally present in the literature: many textbooks postulate more than four rules of quantum mechanics (some up to ten!). This is apart from Nielsen and Chuang, who seem to think that there are only three rules (Nielsen and Chuang 2002).

[3] Especially noteworthy are the attempts to phrase quantum mechanics in an information theoretic way. I recommend the paper by Carlo Rovelli on this subject.

[4] The fact that Dirac split the inner product $\langle|\rangle$—called a "bracket"— into a bra and a ket (the "c" has been lost) sounds a bit like arguing that = sin since the x's can be canceled. Nevertheless, the reader will soon realize that bras and kets are extremely important and useful concepts, and are fully rigorous and justifiable (unlike the cancelation in $\sin x/x$!).

or, in vector notation, as $\mathbf{p} = a\mathbf{x} + b\mathbf{y}$, or, in Dirac notation, as $|p\rangle = a|x\rangle + b|y\rangle$. In the case of Cartesian coordinates, a and b are required to be real numbers. In quantum mechanics, the coefficients a and b are allowed to be complex. A quantum state can be written as $\psi = (\alpha, \beta)$ or, in Dirac notation, $|\psi\rangle = \alpha|x\rangle + \beta|y\rangle$, where α and β are both complex numbers.

A bra represents the complex conjugate transpose of a ket. (The complex conjugate of $x + iy$ is $(x + iy)^* = x - iy$.) If $|\psi\rangle$ is written as a column vector

$$\psi = \begin{pmatrix} \alpha \\ \beta \end{pmatrix}, \tag{2.2}$$

then $\langle\psi|$ is the row vector

$$\psi^\dagger = \begin{pmatrix} \alpha^* & \beta^* \end{pmatrix}, \tag{2.3}$$

where $\psi^\dagger$ is the conjugate transpose of ψ. In Dirac notation $\langle\psi|$ can be written as

$$\langle\psi| = \alpha^*\langle x| + \beta^*\langle y|. \tag{2.4}$$

One of the reasons why bras and kets are useful in Dirac notation is that we often take the product of a vector $|\psi\rangle$ with the complex conjugate transpose of another vector $|\phi\rangle$. The value of $\langle\phi|\psi\rangle$ is called an **inner product**. For example, if $|\psi\rangle = \alpha|x\rangle + \beta|y\rangle$, then the product of $|\psi\rangle$ with its own conjugate transpose represents a quantity known as the **norm** of $|\psi\rangle$, given by

$$\langle\psi||\psi\rangle = \begin{pmatrix} \alpha^* & \beta^* \end{pmatrix} \begin{pmatrix} \alpha \\ \beta \end{pmatrix} \tag{2.5}$$

$$= |\alpha|^2 + |\beta|^2. \tag{2.6}$$

We can work this out using purely bra and ket notation by defining

$$\langle x||y\rangle = \langle y||x\rangle = 0, \tag{2.7}$$
$$\langle x||x\rangle = \langle y||y\rangle = 1. \tag{2.8}$$

These relations mean that the states x and y are orthogonal to each other (i.e. their inner products with each other are equal to zero) and that they are normalized (i.e. that the inner products with themselves are equal to one). Then, to work out $\langle\psi||\psi\rangle$, we multiply out $(\alpha^*\langle x| + \beta^*\langle y|)(\alpha|x\rangle + \beta|y\rangle)$ in a similar way to a standard algebraic expression to obtain

$$\langle\psi||\psi\rangle = (\alpha^*\langle x| + \beta^*\langle y|)(\alpha|x\rangle + \beta|y\rangle) \tag{2.9}$$
$$= \alpha^*\alpha\langle x||x\rangle + \alpha^*\beta\langle x||y\rangle + \beta^*\alpha\langle y||x\rangle + \beta^*\beta\langle y||y\rangle \tag{2.10}$$
$$= \alpha^*\alpha.1 + \alpha^*\beta.0 + \beta^*\alpha.0 + \beta^*\beta.1 \tag{2.11}$$
$$= |\alpha|^2 + |\beta|^2, \tag{2.12}$$

which is the same as what we obtained above in eqn 2.6. For any quantum state, we usually assume that its norm is 1—this is called **normalization**. The reason for this, as we shall see later in this chapter, is that $|\alpha|^2$ and $|\beta|^2$ correspond to probabilities when we apply a measurement to $|\psi\rangle$, and probabilities must be conserved (i.e. add up to unity).

Bras and kets can also be used to define matrices. The matrix

$$A = \begin{pmatrix} a_{xx} & a_{xy} \\ a_{yx} & a_{yy} \end{pmatrix} \tag{2.13}$$

can be represented in Dirac notation with the help of the outer product of bras and kets $|.\rangle\langle.|$ as

$$A = a_{xx}|x\rangle\langle x| + a_{xy}|x\rangle\langle y| + a_{yx}|y\rangle\langle x| + a_{yy}|y\rangle\langle y| \tag{2.14}$$

The same trick used as before in eqn 2.12 can be used to work out the effect of the matrix A on a vector $|\psi\rangle = \alpha|x\rangle + \beta|y\rangle$ using purely Dirac notation:

$$\begin{aligned} A|\psi\rangle &= (a_{xx}|x\rangle\langle x| + a_{xy}|x\rangle\langle y| + a_{yx}|y\rangle\langle x| + a_{yy}|y\rangle\langle y|)(\alpha|x\rangle + \beta|y\rangle) \\ &= a_{xx}|x\rangle\langle x||x\rangle + a_{xy}|x\rangle\langle y||x\rangle + a_{yx}|y\rangle\langle x||x\rangle + a_{yy}|y\rangle\langle y||x\rangle \\ &+ a_{xx}|x\rangle\langle x||y\rangle + a_{xy}|x\rangle\langle y||y\rangle + a_{yx}|y\rangle\langle x||y\rangle + a_{yy}|y\rangle\langle y||y\rangle \\ &= a_{xx}|x\rangle + a_{yx}|x\rangle + a_{xy}|x\rangle + a_{yy}|y\rangle\langle y| \\ &= (a_{xx} + a_{yx})|x\rangle + (a_{xy} + a_{yy})|y\rangle \,. \end{aligned} \tag{2.15}$$

You can check that this is the same outcome as if the column vector ψ had been multiplied by the matrix A.

2.2 The qubit, higher dimensions, and the inner product

So far, we have talked about the quantum state $|\psi\rangle = \alpha|x\rangle + \beta|y\rangle$. It is more conventional in quantum information theory to write $|\psi\rangle$ as $\alpha|0\rangle + \beta|1\rangle$, which is called a qubit (short for "quantum bit"[5]). The labels $|0\rangle$ and $|1\rangle$ are arbitrary orthogonal, normalized states. $|\psi\rangle$ might represent a superposition of the two spins of an electron, $|\psi\rangle = \alpha|\uparrow\rangle + \beta|\downarrow\rangle$ or a photon in a superposition of two different polarizations, $|\psi\rangle = \alpha|H\rangle + \beta|V\rangle$, where $|H\rangle$ and $|V\rangle$ represent the horizontal and vertical polarizations, respectively. When $|\psi\rangle$ is written in this way, it is easy to extend $|\psi\rangle$ to higher dimensions. $|\psi\rangle = \alpha_0|0\rangle + \alpha_1|1\rangle$ is a two-dimensional quantum state. The state

$$\begin{aligned} |\psi\rangle &= \alpha_0|0\rangle + \alpha_1|1\rangle + \ldots + \alpha_{n-1}|n-1\rangle \tag{2.16} \\ &= \sum_{i=0}^{n-1} \alpha_i|i\rangle \tag{2.17} \end{aligned}$$

is an n-dimensional quantum state, which has a column vector representation

$$\psi = \begin{pmatrix} \alpha_0 \\ \vdots \\ \alpha_{n-1} \end{pmatrix} . \tag{2.18}$$

---

[5]This word was invented by Ben Schumacher in 1993 (his paper with the word "qubit" only appeared two years later because Ben could not be bothered to reply to his referees!). I must admit though that once I introduced him as the man who invented qubit, to which he complained that he only "discovered" it—it was really God who invented it.

As in eqns 2.7 and 2.8, we can define
$$\begin{aligned}\langle i||j\rangle &= 0 \quad \text{if } i \neq j\,, \\ \langle i||j\rangle &= 1 \quad \text{if } i = j\,,\end{aligned} \tag{2.19}$$
to be able to work out how vectors behave using purely Dirac notation. Physically, the orthogonality between different kets means that the corresponding physical states which the kets represent can be fully distinguished. The connection between distinguishably and the inner product of states will be discussed in Chapter 3.

An important quantity in relation to two quantum states $|\psi\rangle$ and $|\phi\rangle$ is the inner product $\langle \phi||\psi\rangle$. Using the same processes as in eqn 2.12, we find that the inner product of the two quantum states $|\phi\rangle = \sum_j \beta_j |j\rangle$ and $|\psi\rangle = \sum_i \alpha_i |i\rangle$ (where α_i and β_j are complex numbers unrelated to α and β) is

$$\begin{aligned}\langle\phi||\psi\rangle &= \sum_j \beta_j^* \langle j| \sum_i \alpha_i |i\rangle & (2.20) \\ &= \sum_{i,j} \beta_j^* \alpha_i \langle j||i\rangle & (2.21) \\ &= \sum_i \beta_i^* \alpha_i \langle i||i\rangle + \sum_{i \neq j} \beta_j^* \alpha_i \langle j||i\rangle & (2.22) \\ &= \sum_i \beta_i^* \alpha_i.1 + \sum_{i \neq j} \beta_j^* \alpha_i.0 & (2.23) \\ &= \sum_i \alpha_i \beta_i^* \,. & (2.24)\end{aligned}$$

We shall see later on how the inner product gives a measure of the distance between two vectors $|\psi\rangle$ and $|\phi\rangle$.

2.3 Hilbert spaces

We have already been using Hilbert spaces when we talked about Dirac notation and relied on intuition to perform arithmetic with vectors. We now define formally what a Hilbert space is. A Hilbert space is a complete complex vector space upon which an inner product is defined. ("Complete means" that every Cauchy sequence in the space converges to an element in the space, the reasons for this are beyond the scope of this book).

A complex vector space V is a set which contains an element 0, and, for all $|\psi\rangle, |\phi\rangle, |\chi\rangle \in V$ and for all complex numbers α and β, the following laws hold:

$$\begin{aligned}|\psi\rangle + |\phi\rangle &= |\phi\rangle + |\psi\rangle\,, & (2.25) \\ (|\psi\rangle + |\phi\rangle) + |\chi\rangle &= |\psi\rangle + (|\phi\rangle + |\chi\rangle)\,, & (2.26) \\ 0 + |\psi\rangle &= |\psi\rangle\,, & (2.27) \\ |\psi\rangle - |\psi\rangle &= 0\,, & (2.28) \\ \alpha(\beta|\psi\rangle) &= (\alpha\beta)|\psi\rangle\,, & (2.29) \\ (\alpha + \beta)|\psi\rangle &= \alpha|\psi\rangle + \beta|\psi\rangle\,, & (2.30) \\ \alpha(|\psi\rangle + |\phi\rangle) &= \alpha|\psi\rangle + \alpha|\phi\rangle\,, & (2.31) \\ 1|\psi\rangle &= |\psi\rangle\,. & (2.32)\end{aligned}$$

18 Quantum mechanics

An inner product $\langle .||. \rangle$ is like a dot product and obeys the following rules:

$$(\alpha\langle\psi| + \beta\langle\psi|)|\chi\rangle = \alpha\langle\psi||\chi\rangle + \beta\langle\psi||\chi\rangle \,, \tag{2.33}$$
$$\langle\psi||\phi\rangle = \langle\phi||\psi\rangle^* \,, \tag{2.34}$$
$$\langle\psi||\psi\rangle \geq 0 \text{ and } = 0 \text{ if and only if } |\psi\rangle = 0 \,. \tag{2.35}$$

You can check that the inner product we have been using follows these rules.

Given a Hilbert space H, a spanning set $\{|\phi_i\rangle\}_i$ is a set of vectors such that each $|\psi\rangle \in H$ can be written as a linear combination of $|\phi_i\rangle$'s,

$$|\psi\rangle = \sum_i \alpha_i |\phi_i\rangle \,. \tag{2.36}$$

A set of n vectors $\{|\psi_i\rangle\}_{i=1}^n$ is called **linearly independent** if, for any set of n nonzero complex numbers $\{\alpha_i\}_{i=1}^n$, $\sum_i \alpha_i |\psi_i\rangle = 0$. What linear independence means is that no vector $|\psi_j\rangle$ in a set of linearly independent vectors can be written as a linear combination of the other vectors: $|\psi_j\rangle \neq \sum_{i \neq j} \alpha_i |\psi_i\rangle$ for any α_i's.

A set of states is **orthogonal** if every state is orthogonal to every other state. In this book, we rarely use nonorthogonal linearly independent sets. Actually, all sets of orthogonal states are linearly independent. Let $S = \{|\psi_i\rangle\}_{i=0}^n$ be a set of orthogonal states. Suppose that they are not linearly independent. Then some $|\psi_j\rangle$ can be written as $|\psi_j\rangle = \sum_{i \neq j} \alpha_i |\psi_i\rangle$. By the definition of being a state, the norm of $|\psi_j\rangle$ is $\langle\psi_j||\psi_j\rangle = 1$. However, under the assumption that S is not linearly independent, we can work out the norm of $|\psi_j\rangle$ as follows:

$$\langle\psi_j||\psi_j\rangle = \langle\psi_j|(\Sigma_{i \neq j}\alpha_i|i\rangle) \tag{2.37}$$
$$= \sum_{i \neq j} \alpha_i \langle\psi_j||\psi_i\rangle \tag{2.38}$$
$$= 0 \tag{2.39}$$

as S is an orthogonal set. Therefore $|\psi_j\rangle$ is not a linear combination of the other states in S, which leads us to conclude that any set of orthogonal vectors is linearly independent.

If H is a Hilbert space and S is a set of linearly independent vectors which span H, then S is called a **basis** for H. All bases for H have the same size n; n is called the **dimension** of H. Since orthogonal vectors are always linearly independent, it suffices to find a set of n orthogonal vectors in H to find a basis for H. The sets $\{|0\rangle, |1\rangle\}$ and $\{|0\rangle + |1\rangle, |0\rangle - |1\rangle\}$ are both bases for the space of all qubits[6].

The idea of orthogonal bases can easily be understood by considering Cartesian coordinates. The **xy** coordinates form a two-dimensional vector space and any point $|p\rangle$ in the **xy** plane can be written uniquely as $|p\rangle = a|x\rangle + b|y\rangle$. Any two vectors $|x'\rangle$ and $|y'\rangle$ which are orthogonal are geometrically at right angles and so form an alternative set of axes by which p can be described uniquely as $|p\rangle = a'|x'\rangle + b'|y'\rangle$.

---
[6]The second basis is not normalized—can you show how to normalize it?

2.4 Projective measurements and operations

We have said that the state $|\psi\rangle = \alpha|0\rangle + \beta|1\rangle$ is a possible state of a two-level quantum system, called a qubit. When $\alpha = 1$ and $\beta = 0$, $|\psi\rangle$ takes the value $|0\rangle$, and when $\alpha = 0$ and $\beta = 1$, $|\psi\rangle$ takes the value $|1\rangle$, just like a classical (nonquantum) bit. When $|\psi\rangle$ does not take the value $|0\rangle$ or $|1\rangle$, it is said to be in a superposition of the two values.

A qubit in the state $|\psi\rangle = \alpha|0\rangle + \beta|1\rangle$ can be measured to see if it is in the state $|0\rangle$ or the state $|1\rangle$, and it collapses to the state $|0\rangle$ with probability $|\alpha|^2$ or to the state $|1\rangle$ with probability $|\beta|^2$. To ensure that the probabilities add up to one, we require that $|\alpha|^2 + |\beta|^2 = 1$ or, in Dirac notation, $\langle\psi||\psi\rangle = 1$. Two other useful basis states of qubits are the Hadamard states. These are

$$|+\rangle = \frac{1}{\sqrt{2}}(|0\rangle + |1\rangle), \qquad (2.40)$$

$$|-\rangle = \frac{1}{\sqrt{2}}(|0\rangle - |1\rangle). \qquad (2.41)$$

The Hadamard states $|+\rangle$ and $|-\rangle$ are sometimes called quantum coins as when either one is measured, it collapses to either $|0\rangle$ or $|1\rangle$ with probability 0.5. The Hadamard states are orthogonal, which means that their inner product is zero, that is, $\langle+||-\rangle = 0$. In general, a projective measurement performed on a quantum state $|\psi\rangle$ collapses to $|i\rangle$ with probability $\langle i||\psi\rangle\langle\psi||i\rangle = |\langle i||\psi\rangle|^2$.

An operation O can be represented in Dirac notation as $O = \sum_{i,j} \alpha_{i,j}|i\rangle\langle j|$, where α_{ij} represents the entry in the ith row and the jth column of the matrix form of O.

A projective measurement operator on a state M is represented by an outer product of the form $|x\rangle\langle x|$. This operator is Hermitian meaning that $M = M^\dagger$. It is also called a projection operator or an idempotent operator (meaning that its square is equal to itself, i.e. $|x\rangle\langle x||x\rangle\langle x| = |x\rangle\langle x|$).

An example of the action of a projection operator is a projection of the state $|+\rangle = (|0\rangle + |1\rangle)/\sqrt{2}$ onto the state $|0\rangle\langle 0|$:

$$|0\rangle\langle 0|\frac{1}{\sqrt{2}}(|0\rangle + |1\rangle) = \frac{1}{\sqrt{2}}|0\rangle. \qquad (2.42)$$

The final state is therefore the state $|0\rangle$, and the factor in front, when we take its modulus squared, gives us the probability of the projection being successful (namely one-half in this case). The fact that the modulus squares provides the rule giving us the probability in quantum mechanics is something that is known as the Born postulate (this is our third rule of quantum mechanics, as will be shown later in this chapter). The probability that $|+\rangle$ collapses into the state $|0\rangle$ is given by $1/2 = \langle+||0\rangle\langle 0||+\rangle$.

Hermitian operators play an important role in quantum mechanics because they represent observables, that is, all the quantities that can be measured experimentally. So, positions and momenta of particles can be represented by Hermitian operators, as well as the polarization of a photon or a spin of an electron. Projective measurements are Hermitian operators too.

2.5 Unitary operations

The state of an isolated quantum system, such as a qubit, is in general changed by a unitary operation.[7] A operation U is unitary if it can be written as

$$U = \sum_i |\psi_i\rangle\langle\phi_i|, \qquad (2.43)$$

where the $|\psi_i\rangle$ and $|\phi_i\rangle$ both form orthogonal bases. A simple example of a unitary operation is the identity I,

$$I = \sum_i |\psi_i\rangle\langle\psi_i|, \qquad (2.44)$$

where $|\psi_i\rangle$ is a basis. Since $|\psi_i\rangle$ is a basis, any vector $|\psi\rangle$ can be written as $|\psi\rangle = \sum_i \alpha_i |\psi_i\rangle$. You can easily check that the effect of the identity operation is

$$\begin{aligned} I|\psi\rangle &= I\sum_i \alpha_i |\psi_i\rangle & (2.45) \\ &= \sum_i \alpha_i |\psi_i\rangle\langle\psi_i|\psi_i\rangle & (2.46) \\ &= |\psi\rangle, & (2.47) \end{aligned}$$

which is to leave $|\psi\rangle$ unchanged. The effect of the identity operation is independent of the basis that we choose to represent it.

The inverse of an operation O is the operation O^{-1}, which inverts it so that $O^{-1}O|\psi\rangle = OO^{-1}|\psi\rangle = |\psi\rangle$. In other words,

$$O^{-1}O = OO^{-1} = I \qquad (2.48)$$

. Not all operations have inverses. Unitary operations have (by definition!). A useful property of unitary operations is that the conjugate transpose $U^\dagger$ of a unitary operation U is its inverse:

$$\begin{aligned} UU^\dagger &= \sum_i |\psi_i\rangle\langle\phi_i| \sum_j |\phi_j\rangle\langle\psi_j| & (2.49) \\ &= \sum_{i,j} \langle\phi_i||\phi_j\rangle |\psi_i\rangle\langle\psi_j| & (2.50) \\ &= \sum_i |\psi_i\rangle\langle\psi_i|. & (2.51) \end{aligned}$$

$UU^\dagger$ has the effect that $|\psi_i\rangle$ is transformed into the state $|\psi_i\rangle$ — that is, $UU^\dagger$ is the identity operation I. Because all unitary operations have an inverse, they are said to be reversible (in a sense, we can run them backwards and undo what has been done).

---

[7] Unitary dynamics is fully reversible, which makes the quantum dynamics of an isolated system reversible in the same way as classical Newtonian dynamics. The irreversible element comes in—somewhat loosely speaking—from quantum measurements, which we have addressed before and will address in more detail later.

An example of a unitary operation on a single qubit is the Hadamard operation which is
$$H = |+\rangle\langle 0| + |-\rangle\langle 1| . \tag{2.52}$$
H can be expanded into matrix form as follows:
$$\begin{align}
H &= \frac{1}{\sqrt{2}}(|0\rangle + |1\rangle)\langle 0| + \frac{1}{\sqrt{2}}(|0\rangle - |1\rangle)\langle 1| \tag{2.53}\\
&= \frac{1}{\sqrt{2}}(|0\rangle\langle 0| + |1\rangle\langle 0| + |0\rangle\langle 1| - |1\rangle\langle 1|) \tag{2.54}\\
&= \frac{1}{\sqrt{2}}\begin{pmatrix} 1 & 1 \\ 1 & -1 \end{pmatrix}. \tag{2.55}
\end{align}$$

The effect of the Hadamard operation on the two states $|0\rangle$ and $|1\rangle$ is
$$\begin{align}
H|0\rangle &= |+\rangle , \tag{2.56}\\
H|1\rangle &= |-\rangle . \tag{2.57}
\end{align}$$

By expanding $|+\rangle$ and $|-\rangle$ in terms of $|0\rangle$ and $|1\rangle$, you can see that the effect of the Hadamard operation on the Hadamard states is:
$$\begin{align}
H|+\rangle &= |0\rangle , \tag{2.58}\\
H|-\rangle &= |1\rangle . \tag{2.59}
\end{align}$$

Hadamard states are equally likely to collapse into $|0\rangle$ or $|1\rangle$ when a measurement is performed. However, if a Hadamard operation is performed on a Hadamard state before a measurement is performed, then a measurement can distinguish between the Hadamard states with perfect accuracy. This is one of the strange features of quantum mechanics and is the basis for the quantum weirdness of the Mach–Zehnder interferometer.

2.6 Eigenvectors and eigenvalues

Some important properties of operators are their eigenvectors and eigenvalues. An eigenvector of an operator O is any vector $|e\rangle$ such that
$$O|e\rangle = \lambda|e\rangle , \tag{2.60}$$
and λ is the eigenvalue associated with the eigenvector $|e\rangle$. For example, let $\sigma_1 = |0\rangle\langle 1| + |1\rangle\langle 0|$. Then
$$\sigma_1|+\rangle = |+\rangle . \tag{2.61}$$
So $|+\rangle$ is an eigenvalue of σ_1, with eigenvector 1. The other eigenvector of σ_1 is $|-\rangle$ with eigenvalue -1. You can check that
$$\sigma_1|-\rangle = -|-\rangle . \tag{2.62}$$

In general, if $|e_1\rangle$ and $|e_2\rangle$ are two eigenvectors of an operator O corresponding to the same eigenvalue λ, then

$$O(\alpha|e_1\rangle + \beta|e_2\rangle) = \alpha O|e_1\rangle + \beta O|e_2\rangle \qquad (2.63)$$
$$= \alpha\lambda|e_1\rangle + \beta\lambda|e_2\rangle \qquad (2.64)$$
$$= \lambda(\alpha|e_1\rangle + \beta|e_2\rangle) , \qquad (2.65)$$

and so any linear combination of two eigenvectors with the same eigenvalue is also an eigenvector. This is enough to show that the set of all eigenvectors with the same eigenvalue forms a vector space (given that they are in a vector space, all we have to show is closure). It is easy to see from this that if $|e\rangle$ is an eigenvector with an eigenvalue distinct from those of eigenvectors $|e_1\rangle, \ldots, |e_k\rangle$, then $|e\rangle$ is linearly independent from $|e_1\rangle, \ldots, |e_k\rangle$.

A normal operator N is an operator for which $N^\dagger N = NN^\dagger$. You can check that Hermitian operators and unitary operators are normal — we often deal with normal operators in quantum mechanics. Two eigenvectors of a normal operator with distinct eigenvalues are orthogonal.

The operator σ_1 is a normal operator. We can write σ_1 as a sum of its eigenvalues and eigenvectors as follows:

$$\sigma_1 = |+\rangle\langle+| - |-\rangle\langle-| . \qquad (2.66)$$

In the next section we show that this holds much more generally—any normal operator can be written in this way. From this, we can relate the eigenvalues of an operator to properties of the operator itself.

2.7 Spectral decomposition

The concept of normal operators is very useful because normal operators have a spectral decomposition. That is, if N is a normal operator, then N can be written as

$$N = \sum_i \lambda_i |e_i\rangle\langle e_i| , \qquad (2.67)$$

where the e_i's are orthogonal eigenvectors of N and the λ_i's are their associated eigenvalues. This is called the **diagonal form** —if we were to write N as a matrix in the $|e_i\rangle$ basis then only the diagonal elements would be nonzero.

It is easy to see that any diagonalized operator is normal. If

$$O = \sum_i \lambda_i |e_i\rangle\langle e_i| , \qquad (2.68)$$

then

$$OO^\dagger = \sum_i \lambda_i |e_i\rangle\langle e_i| \sum_j \lambda_j^* |e_j\rangle\langle e_j| \qquad (2.69)$$
$$= \sum_i |\lambda_i|^2 |e_i\rangle\langle e_i| \qquad (2.70)$$
$$= O^\dagger O . \qquad (2.71)$$

The fact that any normal operator can be diagonalized is usually shown by induction on the number of distinct eigenvalues. We shall only summarize the proof here—you can find the actual proof on the Internet or in a book such as Nielsen and Chuang's "Quantum Computation and Quantum Information"

The proof proceeds by induction on the number of eigenvalues. If N is a normal operator with one distinct eigenvalue, then N can be written as

$$N = \sum_i \lambda |e_i\rangle\langle e_i| \qquad (2.72)$$

where the $|e_i\rangle$'s are any orthogonal basis—which establishes the base case.

Now we assume that any normal operator N' can be diagonalized if it has n distinct eigenvalues, and show that this implies that any normal operator N can be diagonalized if it has $n+1$ eigenvalues. Let N be a normal operator with $n+1$ eigenvalues. Then we can write N as

$$\begin{align} N &= INI \qquad &(2.73)\\ &= P_{n+1}NP_{n+1} + P_{1\ldots n}NP_{n+1} + P_{n+1}NP_{1\ldots n} + P_{1\ldots n}NP_{1\ldots n} \qquad &(2.74) \end{align}$$

where P_{n+1} and $P_{1\ldots n}$ are projections onto the spaces spanned by the eigenvectors corresponding to the $(n+1)$th eigenvalue and the first n eigenvalues respectively. The proof proceeds by showing that $P_{n+1}NP_{1\ldots n}$ and $P_{1\ldots n}NP_{n+1}$ are zero and then showing that $P_{1\ldots n}NP_{1\ldots n}$ is normal so that the induction hypothesis can be applied.

2.8 Applications of the spectral theorem

We shall now derive some properties of unitary and Hermitian operators using spectral decomposition. Let U be a unitary operator. You can check, using our definition in section 2.5, that U is unitary if and only if $UU^\dagger = I$. We can write out the spectral decomposition of a unitary operator as follows:

$$U = \sum_j \lambda_j |e_j\rangle\langle e_j|. \qquad (2.75)$$

Then, since the $|e_i\rangle$'s form an orthogonal basis, we have

$$\begin{align} \sum_j |e_j\rangle\langle e_j| &= I \qquad &(2.76)\\ &= UU^\dagger \qquad &(2.77)\\ &= \sum_j |\lambda_j|^2 |e_j\rangle\langle e_j|. \qquad &(2.78) \end{align}$$

So an operator on an n-dimensional space is unitary if and only if it has n orthogonal eigenvectors and each eigenvalue λ_i satisfies $|\lambda_i|^2 = 1$—which means that each eigenvalue is of the form $\lambda_j = e^{i\phi_j}$, where ϕ_j is some angle between 0 and 2π. We can easily write the inverse of U as

$$U^\dagger = \sum_j e^{-i\phi_j} |e_j\rangle\langle e_j|. \qquad (2.79)$$

Another property of unitary operators (which we can obtain without using spectral decomposition) is that unitary operators preserve inner products (in the sense that

24 *Quantum mechanics*

the inner product of $U|\phi\rangle$ and $U|\psi\rangle$ is the same as the inner product of $|\phi\rangle$ and $|\psi\rangle$).[8] This is because, for any two vectors $|\phi\rangle$ and $|\psi\rangle$,

$$\langle\phi|U^\dagger U|\psi\rangle = \langle\phi|I|\psi\rangle \quad (2.80)$$
$$= \langle\phi||\psi\rangle \quad (2.81)$$

We now turn our attention to Hermitian operators. If we write a Hermitian operator as

$$H = \sum_i \lambda_i |e_i\rangle\langle e_i|, \quad (2.82)$$

Then its conjugate transpose is

$$H^\dagger = \sum_i \lambda_i^* |e_i\rangle\langle e_i|. \quad (2.83)$$

Since $H = H^\dagger$, we immediately see that H is Hermitian if and only if its eigenvalues are real.

Another application of eigenvalues is in defining functions of operators. If we have a normal operator N, we can define N^2 as

$$N^2 = N \times N \quad (2.84)$$
$$= \left(\sum_i \lambda_i |e_i\rangle\langle e_i|\right)\left(\sum_j \lambda_j |e_j\rangle\langle e_j|\right) \quad (2.85)$$
$$= \sum_i \lambda_i^2 |e_i\rangle\langle e_i|. \quad (2.86)$$

More generally, if f is some function, we can define $f(N)$ to be:

$$f(N) = \sum_i f(\lambda_i)|e_i\rangle\langle e_i|, \quad (2.87)$$

which gives us a simple way of applying functions to operators.

2.9 Dirac notation shorthands

Dirac notation has been scrawled rapidly over blackboards for many decades and several informal shorthands have been developed, which we shall use throughout the rest of the book.

An inner product $\langle.||.\rangle$ is often simply written as $\langle.|.\rangle$, missing out the extra vertical line.

---

[8] This is a way in which unitary operators can be defined in the first place. The only operators that preserve inner products are unitary and antiunitary, the latter being unphysical for reasons that we shall not go into here.

Two quantum states $|\psi\rangle = \sum_i \alpha_i |i\rangle$ and $|\phi\rangle = \sum_j \alpha_j |j\rangle$ can be combined to form a joint state by taking the tensor product

$$|\psi\rangle \otimes |\phi\rangle = \sum_{i,j} \alpha_i \beta_j |i\rangle \otimes |j\rangle . \tag{2.88}$$

However, it is often more convenient to miss out the $\otimes$, writing the tensor product of two states simply as

$$|\psi\rangle \otimes |\phi\rangle = |\psi\rangle|\phi\rangle = |\psi\phi\rangle . \tag{2.89}$$

Quantum states such as the Hadamard state $|+\rangle$ are formally written as

$$|+\rangle = \frac{1}{\sqrt{2}}(|0\rangle + |1\rangle) . \tag{2.90}$$

However it is very convenient to miss out the normalization factor of $\frac{1}{\sqrt{2}}$ and simply write $|+\rangle$ as

$$|+\rangle = |0\rangle + |1\rangle . \tag{2.91}$$

The normalization factor can be added in at any time that it is needed.

2.10 The Mach–Zehnder interferometer

The Mach–Zehnder interferometer setup is one of many experiments which have convinced physicists that quantum mechanics cannot be described by classical mechanics. The experiment is shown in Fig. 2.1. The initial state of the photon is $|0\rangle$ or $|1\rangle$ (at the position of the light source). How does the beam splitter now act? Well, it sends $|0\rangle$ into $i|2\rangle + |3\rangle$ and sends $|1\rangle$ into $|2\rangle + i|3\rangle$ (2 and 3 label the output ports; the imaginary phase is there because reflection through a right angle generates this phase). The beam splitter executes a unitary transformation. The main property of any unitary transformation is that it sends orthogonal states into orthogonal states. Is this true for the beam splitter? The overlap of the output states is their inner product

$$(-i\langle 2| + \langle 3|)(|2\rangle + i|3\rangle) = -i + i = 0 , \tag{2.92}$$

and so they are indeed orthogonal! What does the operator representing the beam splitter look like? This is it:

$$U_{\text{BS1}} = (i|2\rangle + |3\rangle)\langle 0| + (|2\rangle + i|3\rangle)\langle 1| . \tag{2.93}$$

Let's see what happens when the beam splitter operates on $|1\rangle$. We get

$$\{(i|2\rangle + |3\rangle)\langle 0| + (|2\rangle + i|3\rangle)\langle 1|\}|1\rangle = (i|2\rangle + |3\rangle)\langle 1|1\rangle = (|2\rangle + i|3\rangle) . \tag{2.94}$$

Bingo! We get the right result.

In the next stage of the interferometer, $|2\rangle$ and $|3\rangle$ are both sent via mirrors, picking up another imaginary phase as they are reflected through another right angle. The action of the two mirrors can be written as

$$U_{\text{M1}} = i|5\rangle\langle 2| , \tag{2.95}$$
$$U_{\text{M2}} = i|4\rangle\langle 3| . \tag{2.96}$$

After the action of the two mirrors, the state $|2\rangle + i|3\rangle$ is transformed into $i|5\rangle - |4\rangle$.

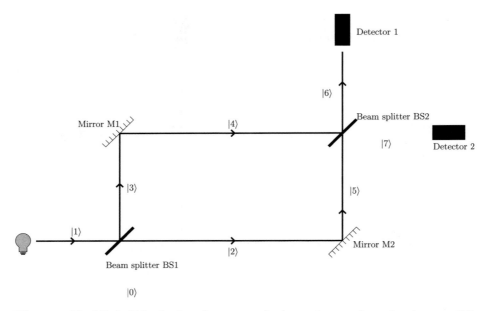

Fig. 2.1 The Mach–Zehnder interferometer. A photon is sent through a beam splitter and bounced off two mirrors into another beam splitter. Mysteriously, only one of the detectors registers a photon.

What is the effect of the second beam splitter on $|4\rangle$ and $|5\rangle$? The effect of the second beam splitter can be written as

$$U_{\text{BS2}} = (i|6\rangle + |7\rangle)\langle 4| + (|6\rangle + i|7\rangle)\langle 5| \ . \tag{2.97}$$

The state $i|5\rangle - |4\rangle$ under the action of the mirror, becomes

$$\begin{align} U_{\text{BS2}}(i|5\rangle - |4\rangle) &= iU_{\text{BS2}}|5\rangle - U_{\text{BS2}}|4\rangle \tag{2.98}\\ &= i(i|6\rangle + |7\rangle) - (|6\rangle + i|7\rangle) \tag{2.99}\\ &= -|6\rangle \ , \tag{2.100} \end{align}$$

which is why, as shown in the diagram, the photons are observed only in Detector 1!

If an absorber is placed at position 5, corresponding to the state $|5\rangle$, as shown in Fig. 2.2, then, after traveling via the mirrors, the state $i|5\rangle - |4\rangle$ is absorbed with probability 0.5 at position 5 (the absorber is a classical object which "measures" the position of the photon). If the photon is not absorbed, it collapses into the state $|4\rangle$ and is transformed by the second beam splitter into the state $(i|6\rangle + |7\rangle)$ which will be detected at either of the two detectors with probability 0.5.

If we compare the results from the two experiments, without any knowledge of quantum mechanics, the results seem very strange. Somehow, the possible presence of a photon at position 5 (when it is not absorbed) prevents a photon at position 4 from reaching Detector 1!

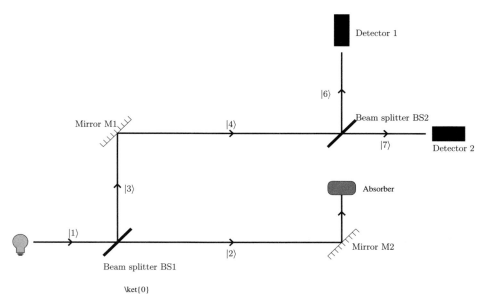

Fig. 2.2 The Mach–Zehnder interferometer with an absorber placed at position $|5\rangle$. The photon is now detected with equal probability in the two detectors.

2.11 The postulates of quantum mechanics

There are four postulates of quantum mechanics which are said to describe all we need to know about quantum systems (i.e. the four postulates can be used to describe any quantum system). These are what you can find in almost all standard textbooks on quantum mechanics. Later on, we shall see that these can and need to be generalized in the light of what we have learned in the field of quantum information (it is always surprising to me to learn that this generalization, although very fundamental, is still unknown to most physicists). The postulates are as follows:

- *States.* States of physical systems are represented by vectors in Hilbert spaces. This postulate says that a physical state in a quantum system can be represented as one of the vectors $|.\rangle$ in the Dirac notation defined above.
- *Observables.* Observables are represented by Hermitian operators. This is because these operators have real eigenvalues, which are appropriate for representing physical quantities (such as an amount of energy, or a distance from the Sun, for example).
- *Measurement.* A quantum state can be measured by use if a set of orthogonal projections. If $|\phi_1\rangle$, ... , $|\phi_k\rangle$ are orthogonal states, then a quantum state $|\psi\rangle$ can be measured by use of $|\phi_1\rangle$, ... , $|\phi_k\rangle$ and collapses into the state $|\phi_i\rangle$ with probability $|\langle\phi_i|\psi\rangle|^2$.
- *Unitary Evolution.* Any change that takes place in a quantum system which is not a measurement can be expressed by the action of a unitary operation.

Measurements are, in general, irreversible processes by which some information is learned about a quantum system. They can be thought of as being irreversible because the information gained from the quantum system is lost from the quantum system itself. The other way in which a quantum system can evolve is reversibly through a unitary operations through which no information about the quantum system is gained or lost.

2.12 Mixed states

The measurement postulate states that a quantum state $|\psi\rangle$ can be measured by use of the orthogonal states $|\phi_1\rangle, \ldots, |\phi_k\rangle$, in which case it collapses to the state $|\phi_i\rangle$ with probability $|\langle\phi_i|\psi\rangle|^2$. Is it possible to write the state of the system using our good old friend Dirac notation without resorting to probability theory? Well, Dirac notation is very versatile, and the answer to that question is, happily, yes.

A probability distribution in which a state $|\phi_i\rangle$ occurs with probability p_i can be written in Dirac notation as a density operator ϱ, given by

$$\varrho = \sum_i p_i |\phi_i\rangle\langle\phi_i| \,. \tag{2.101}$$

The state $|\psi\rangle$ occurs with probability 1 and can be written in density operator notation as $1 * |\psi\rangle\langle\psi| = |\psi\rangle\langle\psi|$; it is called a pure state. We can write the action of the measurement as

$$\sum_i |\phi_i\rangle\langle\phi_i|\psi\rangle\langle\psi|\phi_i\rangle\langle\phi_i| = \sum_i |\langle\phi_i|\psi\rangle|^2 |\phi_i\rangle\langle\phi_i| \,, \tag{2.102}$$

which says that, after the measurement, $|\phi_i\rangle$ occurs with probability $|\langle\phi_i|\psi\rangle|^2$ just as in the measurement postulate.

There are many subtleties associated with density matrices. For example, once you are told the density matrix describing the state of a system, that is all you need to make all possible predictions about the future behavior of the system. There are infinitely many ways of writing down one and the same density matrix, but all of them are completely equivalent as far as the outcomes of measurements on that system are concerned. So, for example, an equal mixture of $|0\rangle$ and $|1\rangle$,

$$\varrho = \frac{1}{2}(|0\rangle\langle 0| + |1\rangle\langle 1|) \,, \tag{2.103}$$

can equally well be written as

$$\varrho = \frac{1}{2}(|+\rangle\langle+| + |-\rangle\langle-|) \,, \tag{2.104}$$

where $|\pm\rangle = (|0\rangle \pm |1\rangle)/\sqrt{2}$ (check it!). And no experimental procedure can distinguish between the two! This notion will take some time to get used to, but since we shall repeatedly come back to it, it will hopefully become second nature to you by the end.

The density operator ϱ of a mixture of states has positive eigenvalues—all the probabilities are positive. Furthermore, it is easy to see that $\varrho = \varrho^\dagger$, and so density operators are Hermitian and hence normal. This means that we can diagonalize them. Given any mixture of states, we can write the density operator of the mixture as

$$\varrho = \sum_i p_i |i\rangle\langle i|, \qquad (2.105)$$

where the p_i's are positive eigenvalues and the $|i\rangle$'s are orthogonal. This is exactly how we would write a classical probability distribution where $P(X = i) = p_i$. The two probability distributions are indistinguishable, even though one is quantum and the other is classical. This is one of the reasons why classical information is often seen as a subset of quantum information. In the following chapters, we shall see that this relationship between mixtures of quantum states and classical probability distributions leads to many similarities between classical and quantum information theory.

2.13 Entanglement

Before wrapping up this chapter, I would like to introduce a mysterious property of quantum systems, entanglement. Entanglement is a property of two or more quantum systems which exhibit correlations that cannot be explained by classical physics. It is a key resource in quantum computation and quantum information theory. A large part of the later chapters of this book will be about understanding entanglement—as this is one of the most exciting issues in the field—but for now I shall only say a few words about it for the sake of completeness of this chapter.

States of two quantum systems can be considered together by taking their tensor product. For example, the two Hadamard states $|+\rangle$ and $|-\rangle$ can be considered together in the form

$$|+\rangle|-\rangle = (|0\rangle + |1\rangle)(|0\rangle - |1\rangle) = |00\rangle + |01\rangle + |10\rangle - |11\rangle. \qquad (2.106)$$

Some well-known examples of entangled states are the Bell states $|\Phi^+\rangle$ and $|\Phi^-\rangle$, defined by

$$|\Phi^+\rangle = |00\rangle + |11\rangle, \qquad (2.107)$$
$$|\Phi^-\rangle = |00\rangle - |11\rangle \qquad (2.108)$$

(we have missed out the $1/\sqrt{2}$ normalization factor). Let's try to express $|\Phi^+\rangle$ as a combination of two qubits $\alpha|0\rangle + \beta|1\rangle$ and $\gamma|0\rangle + \delta|1\rangle$. We expand out to find the values of α, β, γ, and δ:

$$|00\rangle + |11\rangle = |\Phi^+\rangle = (\alpha|0\rangle + \beta|1\rangle)(\gamma|0\rangle + \delta|1\rangle) = \alpha\gamma|00\rangle + \beta\gamma|10\rangle + \alpha\delta|01\rangle + \beta\delta|11\rangle \qquad (2.109)$$

Equating the appropriate parts, we get $\alpha\gamma = \beta\delta = 1$ and $\beta\gamma = \alpha\delta = 0$. The result $\beta\gamma = 0$ implies that either β or γ is zero; however, this cannot be the case, since $\alpha\gamma = \beta\delta = 1$. Therefore we conclude that the Bell state $|\Phi^+\rangle$ cannot be broken down into two separate qubit states, and so we say that $|\Phi^+\rangle$ is entangled.

More generally, we say that a pure quantum state is entangled across two or more systems when it cannot be expressed as a tensor product of states in those systems (the definition is more complex for mixed states, and we shall review it in great detail later). Entanglement can exist between two systems which are spatially separated by a great distance (it is in fact completely independent of the distance, at least in principle). We shall see in Chapter 4 how entanglement can be used to help Alice and Bob communicate using quantum states, and even how Alice can teleport a quantum state to Bob by only sending classical bits.

2.14 Summary

States of physical systems are represented by vectors in quantum mechanics. The Dirac notation is made up of bras $\langle .|$s and kets $|.\rangle$s which together form an inner products, or brakets $\langle .||.\rangle$. The kets are column vectors, and the bras are their conjugate transposes.

The braket notation is also convenient for representing observables in quantum mechanics. Observables are, in turn, represented by Hermitian operators: these have orthogonal eigenstates and real eigenvalues (spectral decomposition). Operators are made up using outer products of bras and kets, $|.\rangle\langle .|$. Projective measurements are also represented by Hermitian operators, which are idempotent (they square to an identity matrix).

When a measurement is made using a projector P on the state $|\psi\rangle$, the probability of the outcome is given by the Born rule, $\langle \psi|P|\psi\rangle$.

Finally, when no measurements are made, the system evolves according to a (reversible) unitary transformation. The Schrödinger equation, which is the most fundamental equation in nonrelativistic quantum mechanics, is in fact a generator of unitary transformations.[9]

---

[9] For readers interested in the details of quantum dynamics I recommend my book "Modern Foundations of Quantum Optics", Imperial College Press (2005).

3
Quantum information—the basics

Information is often thought of as an abstract quantity which has nothing to do with the physical world. However, information reaches us through interaction with the outside world and is stored and processed in our bodies, which are—as far as we know—fully subject to the laws of physics. We can draw two crucial insights from this:

- Information must be encoded into a physical system.
- Information must be processed using physical (dynamical) laws.

These two realizations, usually attributed to Landauer in one form or another, imply that all limitations on information processing follow from the restrictions of the underlying physical laws.[1] We studied information theory in Chapter 1 assuming that information processing is governed by the laws of classical physics (in fact, we never even really needed to make this assumption—it is just implicitly there in the definition of bits). Quantum mechanics is a more accurate description of the microscopic world than classical mechanics, and quantum information, which is governed by the laws of quantum mechanics, is a more accurate description of information theory. The quantum laws of physics are fundamentally different from the classical laws and so, therefore, is the resulting information processing. In one limit, which will be discussed extensively later in this book, quantum information reduces to classical information; otherwise, quantum information is a much more general concept that allows information-processing protocols that have no classical analogue whatsoever. Therefore Shannon's revolutionary theory described in the first chapter will be seen only as a very special case of the theory presented henceforth.

3.1 No cloning of quantum bits

Rather than going into the full axioms and theorems of quantum information straightaway, we shall first talk about some basic differences between bits and qubits. These examples will then be generalized to the quantum equivalents of Shannon's theorems. But quantum information will offer us many more possibilities than this!

The first striking difference between quantum and classical information storage is that we cannot clone "unknown" quantum states. That is, given a quantum state

---

[1] Hence my earlier claim that computation and information theory are branches of physics, and not mathematics. Anyway, one day the whole of mathematics will probably become a branch of physics if we continue with this trend into the future.

such as a qubit $|\psi\rangle = \alpha|0\rangle + \beta|1\rangle$, it is impossible to produce two copies $|\psi\rangle|\psi\rangle$ of $|\psi\rangle$ without knowing the values of α and β. This is due to the linearity of quantum mechanics.

Suppose that U is a unitary operation which clones any qubit $|\psi\rangle$. Then U can clone the states $|0\rangle$ and $|1\rangle$:

$$U|0\rangle = |00\rangle, \qquad (3.1)$$
$$U|1\rangle = |11\rangle. \qquad (3.2)$$

Now, if we apply U to a general qubit $|\psi\rangle = \alpha|0\rangle + \beta|1\rangle$, remembering that a unitary operation can be represented by a matrix and is therefore linear, we get $U|\psi\rangle = \alpha U|0\rangle + \beta U|1\rangle$. Plugging in eqns 3.1 and 3.2, we get

$$U(|\psi\rangle) = \alpha|00\rangle + \beta|11\rangle. \qquad (3.3)$$

However, if U was able to clone $|\psi\rangle$, then we would get

$$U(|\psi\rangle) = |\psi\rangle|\psi\rangle = \alpha^2|00\rangle + \alpha\beta|01\rangle + \alpha\beta|10\rangle + \beta^2|11\rangle \qquad (3.4)$$

Equations 3.3 and 3.4 do not match up, which shows that cloning of quantum states is impossible. Note that this conclusion holds even if we try to involve extra ancillary systems, simply because the overall transformation is linear.

"No cloning" is a powerful concept in quantum mechanics.[2] If we could clone an unknown quantum state then we could have an unlimited number of copies of that state upon which to perform experiments and deduce what the state was. "No cloning" implies that there is a limited amount of information which we can learn about an unknown quantum state. This is in stark contrast to classical physics: we can always learn everything about a classical system even if we have only one copy of it, simply because classical measurements are noninvasive and do not destroy the state of the measured system. For example, when we photocopy a piece of paper, we do not destroy the content of the writing on it.

The no-cloning theorem has many applications. It prevents us from using classical error correction techniques when performing quantum computations—we cannot create backup copies of a state in the middle of a quantum computation. Error correction is vital in quantum computing to counteract the interaction of quantum computations with their environment. We shall see at the end of the book how quantum error correction can be performed.

"No cloning", loosely speaking, also prevents instantaneous communication via quantum entanglement. If Alice and Bob shared an entangled Bell state $|\Phi^+_{AB}\rangle = (|00\rangle + |11\rangle)/\sqrt{2}$, then Alice could transmit a bit using measurements. To transmit a $|0\rangle$, she would measure the Bell state in the standard basis $\{|0\rangle, |1\rangle\}$, and to transmit a $|1\rangle$ she would measure in the Hadamard basis $\{|+\rangle, |-\rangle\}$. If Bob had an unlimited number of copies of his part of the Bell state, he could work out which basis Alice measured in—the basis which always returns the same result when he makes a

---

[2] It may also have wider implications for, say, biological cloning—who knows? The no-cloning theorem was, in fact, used by Wigner to claim that quantum mechanics contradicts the basic tenets of evolution!

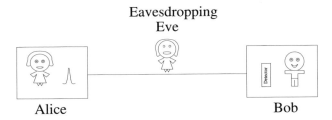

Fig. 3.1 Alice and Bob are trying to communicate secretly, but there is an eavesdropper called Eve who is attempting to listen in on their communications.

measurement. We shall see at the end of the next chapter how instantaneous communication via quantum entanglement is in general impossible (although nonrelativistic quantum mechanics, of course, allows transmission at superluminal velocities—simply by being nonrelativistic!).

3.2 Quantum cryptography

Cryptography is a process by which two parties, Alice and Bob, can communicate secretly. "No cloning" is a vital ingredient in quantum cryptography as it prevents an eavesdropper from copying the quantum states that Alice and Bob send to one another.[3]

The main issue in cryptography is how to establish a secret key between Alice and Bob. This is a string of zeros and ones which is in the possession of both parties, but is not known to any other unwanted parties—that is, eavesdroppers (see Fig. 3.1). Once Alice and Bob have a secret key, they can use it to communicate secretly and exchange any other messages. Establishing this secret key is—as far as we know—impossible to do in an unconditionally safe way within the laws of classical physics.

In the BB84 quantum key exchange cryptography protocol (named after Bennett and Brassard, who came up with this idea in 1984), Alice uses one of four nonorthogonal states to communicate a zero or a one to Bob. The advantage of using nonorthogonal states is that an eavesdropper (usually called Eve) cannot discriminate between them perfectly. Neither can Bob, of course! So how can Alice and Bob communicate without any interception from Eve? We now describe the first such proposal. This will give us the main general idea about how and why quantum cryptography can be successful.

Alice sends Bob photons in one of four polarizations. For example, she may prepare $|+\rangle$, $|-\rangle$. Bob now chooses a measurement basis. The two Hadamard states $|+\rangle$ and $|-\rangle$ have equal probabilities of being collapsed to a $|0\rangle$ or a $|1\rangle$ when measured in the $\{|0\rangle, |1\rangle\}$ basis. Likewise, the two states $|0\rangle$ and $|1\rangle$ have equal chances of being measured so that they collapse to a $|+\rangle$ or a $|-\rangle$ when measured in the $\{|+\rangle, |-\rangle\}$ basis. The idea behind the BB84 protocol is that a bit x_i can be stored either as a quantum bit $|\psi_i\rangle$ in the rectilinear basis as $|0\rangle$ or $|1\rangle$, or in the diagonal basis as $|+\rangle$ or $|-\rangle$ (see Fig. 3.2). If the basis for measurement is chosen wrongly by the eavesdropper Eve,

---

[3]Since cloning is classically possible, there is no unconditional security of classical communication in cryptography.

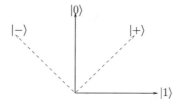

Fig. 3.2 Alice can store x_i either in the rectilinear basis as $|0\rangle$ or $|1\rangle$ or in the diagonal basis as $|+\rangle$ or $|-\rangle$.

there is a 50-50 chance that she will collapse $|\psi_i\rangle$ into the wrong basis, in which case, when it is measured in the correct basis there is a 50-50 chance that it will collapse to the wrong value and her eavesdropping will be detected.

The BB84 protocol begins with Alice choosing a random string $x_1 \ldots x_n$ of bits to send to Bob. For example,

$$\begin{array}{|c|c|c|c|c|} \hline \text{Bit} & x_1 & x_2 & x_3 & x_4 \\ \hline \text{Value} & 0 & 1 & 1 & 0 \\ \hline \end{array}. \tag{3.5}$$

In order to prevent an eavesdropper from reading the bits, Alice randomly chooses to write each bit x_i as a qubit $|\psi_i\rangle$ in either the rectilinear basis as $|0\rangle$ or $|1\rangle$ or in the diagonal basis as $|+\rangle$ or $|-\rangle$. For example,

$$\begin{array}{|c|c|c|c|c|} \hline \text{Classical value} & 0 & 1 & 1 & 0 \\ \text{Alice's basis} & + & \times & + & \times \\ \text{Quantum encoding} & |\psi_1\rangle = |0\rangle & |\psi_2\rangle = |-\rangle & |\psi_3\rangle = |1\rangle & |\psi_4\rangle = |+\rangle \\ \hline \end{array}. \tag{3.6}$$

A logical "zero" is encoded either as $|0\rangle$ or $|+\rangle$, while a logical "one" is encoded as $|1\rangle$ or $|-\rangle$. Alice then sends the encoded string $|\psi_1\rangle \ldots |\psi_n\rangle$ to Bob. When Bob receives the encoded string $|\psi_1\rangle \ldots |\psi_n\rangle$ from Alice, he randomly measures each of the encoded $|\psi_i\rangle$'s in the rectilinear or the diagonal basis. For example,

$$\begin{array}{|c|c|c|c|c|} \hline \text{Classical value} & 0 & 1 & 1 & 0 \\ \text{Alice's basis} & + & \times & + & \times \\ \text{Bob's basis} & \times & \times & + & + \\ \text{In agreement} & \text{No} & \text{Yes} & \text{Yes} & \text{No} \\ \hline \end{array}. \tag{3.7}$$

He then talks to Alice over a public channel (an eavesdropper can listen in if she likes) and they work out which of the qubits $|\psi_i\rangle$ Bob has measured in the same basis as Alice, and discard the others. If there has been no eavesdropping, then Alice and Bob have the same values of x_i for the bits for which the preparation and measurement bases agree. In our example, Alice and Bob both have $|\psi_2\rangle$ and $|\psi_3\rangle$ as $|-\rangle$ and $|1\rangle$ respectively, representing classical bits 0 and 1.

Suppose there is an eavesdropper Eve. She intercepts the message $|\psi_1\rangle \ldots |\psi_n\rangle$ and attempts to make measurements on it to try to discover the value of some of the bits x_i which Alice and Bob are trying to share secretly. For each qubit for which Alice and Bob's measurement bases agree, Eve has a 50-50 chance of guessing incorrectly whether the rectilinear or the diagonal basis is being used, which leads to a 50-50 chance that Alice and Bob's measurements do not agree—a probability of 0.25 overall that Eve's measurement leads to Alice and Bob obtaining different values of x_i. Alice and Bob can randomly pick some of the qubits $|\psi_i\rangle$ for which their bases agree and check that their values for x_i are the same. If they are the same for a large number of qubits, Alice and Bob can be confident that there was no eavesdropping and they can use the remaining qubits for which their bases agree as a secure key. A perfectly secure key exchange between communicating parties is thus possible in quantum mechanics. This is not possible within the laws of classical physics.

3.3 The trace and partial-trace operations

We now review some more theory of quantum mechanics. The trace operation is a useful operation when dealing with density operators. The trace of a matrix is the sum of the diagonal elements. The trace of an operator ϱ, in Dirac notation, is

$$\mathrm{tr}(\varrho) = \sum_k \langle k|\varrho|k\rangle \tag{3.8}$$

and is independent of the choice of the basis $|k\rangle$. The trace of a pure state $|\psi\rangle = \sum_i \alpha_i |i\rangle$ can be worked out by writing it in density operator form:

$$\mathrm{tr}(|\psi\rangle\langle\psi|) = \sum_k \langle k| \sum_{i,j} \alpha_i |i\rangle \alpha_j^* \langle j| |k\rangle \tag{3.9}$$

$$= \sum_{i,j,k} \alpha_i \alpha_j^* \langle k|i\rangle \langle j|k\rangle \tag{3.10}$$

$$= \sum_k |\alpha_k|^2 \tag{3.11}$$

$$= 1. \tag{3.12}$$

The trace of a pure state is always 1 if the state is normalized. The trace of a density operator $\varrho = \sum_i p_i |\psi_i\rangle\langle\psi_i|$ is also always 1 when all the states and corresponding probabilities are normalized:

$$\mathrm{tr}(\varrho) = \sum_k \langle k| \sum_i p_i |\psi_i\rangle\langle\psi_i| |k\rangle \tag{3.13}$$

$$= \sum_i p_i \, \mathrm{tr}(|\psi_i\rangle\langle\psi_i|) \tag{3.14}$$

$$= \sum_i p_i \tag{3.15}$$

$$= 1. \tag{3.16}$$

The fact that this trace is equal to 1 just means that all the probabilities for various outcomes add up to one (as they should!). In the previous chapter, we said that the normalization condition is encoded in the squared modulus of the amplitude for a pure state, and the probabilities for a mixed state sum to 1. We can now use the trace operator to state these normalization conditions more succinctly—$\mathrm{tr}(|\psi\rangle\langle\psi|) =$

$\mathrm{tr}(\varrho) = 1$. We can also use the trace operator to distinguish between pure and mixed states. First we need to take the square of a density matrix:

$$\varrho^2 = \varrho * \varrho \qquad (3.17)$$

$$= \sum_i p_i |\psi_i\rangle\langle\psi_i| \sum_j p_j |\psi_j\rangle\langle\psi_j| \qquad (3.18)$$

$$= \sum_i p_i p_j |\psi_i\rangle\langle\psi_i|\psi_j\rangle\langle\psi_j| \qquad (3.19)$$

$$= \sum_i p_i^2 |\psi_i\rangle\langle\psi_i|. \qquad (3.20)$$

The trace of the density operator squared is then given by

$$\mathrm{tr}(\varrho^2) = \sum_k \langle k | \sum_i p_i |\psi_i\rangle\langle\psi_i||k\rangle \qquad (3.21)$$

$$= \sum_i p_i \, \mathrm{tr}(|\psi_i\rangle\langle\psi_i|) \qquad (3.22)$$

$$= \sum_i p_i^2. \qquad (3.23)$$

If the density operator ϱ represents a pure state, then $\sum_i p_i^2 = 1$. If ϱ is a nontrivial probability distribution of states, then $\mathrm{tr}(\varrho^2) < 1$.

Another important property of the trace operator, which I leave as an exercise for you to prove is that the trace operator is linear — that is, if ϱ and σ are density operators and λ is a complex number, then $\mathrm{tr}(\varrho + \lambda\sigma) = \mathrm{tr}(\varrho) + \lambda \, \mathrm{tr}(\sigma)$. (Hint: a simple proof is obtained by just considering the trace operator as the sum of the diagonal elements of a matrix.)

A very convenient way of thinking about the trace is to say that this operation converts matrices into numbers. In quantum mechanics, observables and states are, as we have seen, represented by matrices. However, in reality we never observe matrices— this is just plain impossible! What we observe are numbers, which represent outcomes of our measurements. The trace is an operation which converts operators (matrices) into numbers, that is,

$$\mathrm{tr}|\psi\rangle\langle\phi| = \langle\psi|\phi\rangle, \qquad (3.24)$$

since the quantity on the right-hand side is in general a complex number. Of course, we do not measure complex numbers either in the real world, but the squared modulus of this number, $|\langle\psi|\phi\rangle|^2$, is the probability of measuring state ψ if we are in the state ϕ, or vice versa.

A very useful application of the trace operator is the partial trace. A partial trace can be used when we have two or more quantum systems but we would like to know only what happens in the first system. The other systems are never ever measured or used (say, we have thrown them into a black hole and no information can ever be recovered about them). Imagine Alice and Bob are two parties who share a mixture of quantum states $\varrho_A \otimes \varrho_B = \sum_{ij} p_{ij} |i_A\rangle \otimes |j_B\rangle\langle i_A| \otimes \langle j_B|$, where the A and B indicate the parts of the system which belong to Alice and Bob, respectively; then the partial-trace operator $\text{tr}_B \varrho^A \otimes \sigma^B = \varrho^A \otimes \text{tr}(\sigma^B)$ can be used to obtain the density operator for Alice:

$$\text{tr}_B(\varrho^A \otimes \sigma^B) = \text{tr}_B \left(\sum_{ij} p_{ij} |i_A\rangle \otimes |j_B\rangle \langle i_A| \otimes \langle j_B| \right) \quad (3.25)$$

$$= \sum_{i,j} |i_A\rangle\langle i_A| \otimes \text{tr}(p_{ij}|j_B\rangle\langle j_B|) \quad (3.26)$$

$$= \sum_{i,j} p_{i,j} |i_A\rangle\langle i_A| = \sum_i p_i |i_A\rangle\langle i_A| \quad (3.27)$$

(where we have defined $p_i = \sum_j p_{ij}$). The last equality follows from the fact that summing over j must give us unity. The physical motivation for using the partial trace is the following. Suppose that we wish to predict measurements on or the evolution of a system (e.g. an atom), but we know that this system is entangled with some other systems (say other atoms). In general we do not want to know what the other systems are doing (frequently this is far beyond our experimental abilities). This then means that we should trace them out from the overall state.

This formula happens to be true even when the joint state of the systems A and B is more complicated, that is, if it is entangled. Suppose that the joint state is $|\psi_{AB}\rangle = \sum_i \alpha_i |i_A\rangle|i_B\rangle$. The density matrix of A and B is then

$$\varrho_{AB} = |\psi_{AB}\rangle\langle\psi_{AB}| = \sum_i \alpha_i |i_A\rangle|i_B\rangle \sum_m \alpha_j^* \langle j_A|\langle j_B| . \quad (3.28)$$

Tracing out the system B now gives us

$$\text{tr}_B \varrho_{AB} = \sum_{i,j} \alpha_i \alpha_j^* |i_A\rangle\langle j_A|\langle j_B||i_B\rangle \quad (3.29)$$

$$= \sum_{i,j} \alpha_i \alpha_j^* |i_A\rangle\langle j_A|\delta_{ij} \quad (3.30)$$

$$= \sum_i |\alpha_i|^2 |i_A\rangle\langle i_A| . \quad (3.31)$$

Note that this state contains only the diagonal elements and the corresponding probabilities in the basis $|i\rangle$. Tracing over one of the two entangled systems, loosely speaking, kills the off-diagonal elements of the other subsystems. This is sometimes taken to be equivalent to the emergence of classicality (i.e. no superpositions of states with different i's) in a subsystem through entanglement with another system.

In the same way that a system can be traced out from a total state, we can actually perform the reverse action—we can add another system to an already existing system. This extension of the total Hilbert space will be an extremely useful technique for us in quantum information theory.

3.4 Hilbert space extension

Suppose Alice and Bob share a Bell State

$$|\Phi^+\rangle = \frac{1}{\sqrt{2}}(|0_A 0_B\rangle + |1_A 1_B\rangle). \tag{3.32}$$

Alice's part of the system can be obtained by tracing out Bob's:

$$
\begin{aligned}
|\Phi_A^+\rangle\langle\Phi_A^+| &= \text{tr}_B\left(\frac{1}{2}(|0_A 0_B\rangle + |1_A 1_B\rangle)(\langle 0_A 0_B| + \langle 1_A 1_B|)\right) & (3.33) \\
&= \frac{1}{2}(|0_A\rangle\langle 0_A| \otimes \text{tr}(|0_B\rangle\langle 0_B|) + |1_A\rangle\langle 0_A| \otimes \text{tr}(|1_B\rangle\langle 0_B|) & (3.34) \\
&\quad + |0_A\rangle\langle 1_A| \otimes \text{tr}(|0_B\rangle\langle 1_B|) + |1_A\rangle\langle 1_A| \otimes \text{tr}(|1_B\rangle\langle 1_B|)) & (3.35) \\
&= \frac{1}{2}|0\rangle\langle 0| + \frac{1}{2}|0\rangle\langle 0|. & (3.36)
\end{aligned}
$$

Alice and Bob share a pure quantum state, though Alice's part of the state is a mixture. In fact any mixture can be considered as part of a pure state.

A pure state which represents a mixture is called a **purification**. One purification of a mixture with a density operator $\varrho = \sum_i p_i |i\rangle\langle i|$ is simply $|\psi\rangle = \sum_i \sqrt{p_i}|i_A\rangle \otimes |i_B\rangle$ (there are infinitely many possible purifications in general). The state ϱ can be obtained from $|\psi\rangle$ by tracing out Bob's part of the system: $\varrho = \text{tr}_B(|\psi\rangle\langle\psi|)$.

3.5 The Schmidt decomposition

If $|\psi^{AB}\rangle$ is a pure bipartite state (a state consisting of two parts, A and B), and $\{|i^A\rangle\}$ and $\{|j^B\rangle\}$ are bases for the systems A and B, respectively, then $|\psi^{AB}\rangle$ can be written as

$$|\psi^{AB}\rangle = \sum_{i,j} \beta_{ij}|i^A\rangle|j^B\rangle \tag{3.37}$$

for some complex numbers β_{ij}. The Schmidt decomposition finds two bases $\{|\psi_i^A\rangle\}$ and $\{|\phi_j^B\rangle\}$ such that $|\psi^{AB}\rangle$ can be written as

$$|\psi^{AB}\rangle = \sum_i \alpha_i |\psi_i^A\rangle|\phi_i^B\rangle, \tag{3.38}$$

which is a much more convenient form that sums over only the i's rather than the i's and j's. System A or B can be traced out immediately from the result of the Schmidt decomposition to obtain

$$
\begin{aligned}
\text{tr}_A(|\psi^{AB}\rangle) &= \sum_i |\alpha_i|^2 |\psi_i^A\rangle\langle\psi_i^A|, & (3.39) \\
\text{tr}_B(|\psi^{AB}\rangle) &= \sum_i |\alpha_i|^2 |\phi_i^B\rangle\langle\phi_i^B|, & (3.40)
\end{aligned}
$$

from which we can find many properties, such as whether the two systems are entangled (to be discussed later at great length). Note that the Schmidt decomposition can only be performed on pure states.

Suppose that A is an m-dimensional system and B is an n-dimensional system, where $m > n$ (we do not lose generality, since the two systems are interchangeable). We can construct the Schmidt decomposition by the following steps:

1. First we write $|\psi^{AB}\rangle$ in density operator form as

$$\varrho^{AB} = |\psi^{AB}\rangle\langle\psi^{AB}| \qquad (3.41)$$

$$= \sum_{i,j}\alpha_{ij}|i^Aj^B\rangle\sum_{k,l}\alpha_{kl}^*\langle k^Al^B| \qquad (3.42)$$

$$= \sum_{i,j,k,l}\alpha_{ij}\alpha_{kl}^*|i^A\rangle\langle k^A|\otimes|j^B\rangle\langle l^B|. \qquad (3.43)$$

2. We trace out system B to obtain

$$\varrho^A = \mathrm{tr}_B(\varrho^{AB}) \qquad (3.44)$$

$$= \sum_r\sum_{i,j,k,l}\alpha_{ij}\alpha_{kl}^*|i\rangle\langle k|\otimes\langle r|j\rangle\langle l|r\rangle \qquad (3.45)$$

$$= \sum_{i,j,r}\alpha_{ir}\alpha_{kr}^*|i\rangle\langle k|. \qquad (3.46)$$

3. We diagonalize ϱ^A to obtain

$$\varrho^A = \sum_i|\beta_i|^2|\psi_i\rangle\langle\psi_i|. \qquad (3.47)$$

4. We return to the joint system and reexpress $|\psi^{AB}\rangle$ in terms of $|\psi_i^A\rangle$ as

$$|\psi^{AB}\rangle = \sum_{i,j}c_{ij}|\psi_i^A\rangle|j^B\rangle, \qquad (3.48)$$

where $c_{ij} = \langle\psi_i^A j^B|\psi^{AB}\rangle$.

5. To get system B in Schmidt form, we define $|\phi_i\rangle$ to be

$$|\phi_i\rangle = \sum_j\frac{c_{ij}}{\beta_i}|j\rangle. \qquad (3.49)$$

The $|\phi_i\rangle$'s are orthogonal, since

$$\langle\phi_i|\phi_j\rangle = \sum_{k,l}\frac{c_{ik}^*}{\beta_i}\langle k|l\rangle\frac{c_{jl}}{\beta_j} \qquad (3.50)$$

$$= \sum_k\frac{c_{ik}^*}{\beta_i}\frac{c_{jk}}{\beta_j} \qquad (3.51)$$

$$= \frac{1}{\beta_i\beta_j}\sum_k\langle\psi^{AB}|\psi_i^A k^B\rangle\langle\psi_j^A k^B|\psi^{AB}\rangle \qquad (3.52)$$

$$= \frac{1}{\beta_i\beta_j}\langle\psi_i^A|\varrho^A|\psi_j^A\rangle \qquad (3.53)$$

$$= \frac{1}{\beta_i\beta_j}\langle\psi_i^A|\sum_k|\beta_i|^2|\psi_k\rangle\langle\psi_k||\psi_j^A\rangle \qquad (3.54)$$

$$= 0. \qquad (3.55)$$

40 *Quantum information—the basics*

6. We can now rewrite $|\psi^{AB}\rangle$ in terms of $|\psi_i^A\rangle$ and $|\psi_i^B\rangle$ as

$$|\psi^{AB}\rangle = \sum_{i,j} c_{ij}|\psi_i^A\rangle|j^B\rangle \tag{3.56}$$

$$= \sum_{i,j} \beta_i |\psi_i^A\rangle \frac{c_{ij}}{\beta_i}|j^B\rangle \tag{3.57}$$

$$= \sum_i \beta_i |\psi_i\rangle|\phi_i\rangle , \tag{3.58}$$

which is in Schmidt form.

The Schmidt decomposition will be very important when we discuss entanglement between two subsystems. We can already make an important statement: if we trace out either part A or part B, the density operator of the remaining part has the same eigenvalues, that is, the two parts are equally mixed.

3.6 Generalized measurements

In the previous chapter, we described a measurement as a set of orthogonal projections, however an auxiliary system can be used to make nonorthogonal measurements. If M_1, ..., M_k are any set of operators which satisfy the completeness relation

$$\sum_i M_i^\dagger M_i = I , \tag{3.59}$$

they are called a **completely positive map** (or **CP-map**) and can be used in conjunction with an auxiliary system to define measurements.

If $M_1, \ldots, M_k$ are a set of operators that satisfy the completeness relation, they can be made into a unitary operation as follows:

$$U|\psi\rangle \otimes |0\rangle = \sum_i M_i|\psi\rangle \otimes |i\rangle . \tag{3.60}$$

A nonorthogonal measurement is made by measuring the auxiliary system in the $\{|i\rangle\}$ basis to obtain $|i\rangle$, with a probability

$$P(i) = \langle\psi|M_i^\dagger M_i|\psi\rangle = \text{tr}(M_i|\psi\rangle\langle\psi|M_i^\dagger) . \tag{3.61}$$

If i is the outcome of the measurement, then the nonauxiliary system is left in the state

$$\frac{M_i|\psi\rangle}{\sqrt{P(i)}} . \tag{3.62}$$

The completeness relation also ensures that U is unitary. If $|\psi\rangle$ and $|\phi\rangle$ are two states, then U preserves their inner product:

$$(\langle\phi| \otimes \langle 0|U^\dagger)(U|\psi\rangle \otimes |0\rangle) = \sum_{i,j} \langle\phi|M_i^\dagger \otimes \langle i|M_j|\psi\rangle \otimes |j\rangle \tag{3.63}$$

$$= \sum_{i,j} \langle\phi|M_i^\dagger M_j|\psi\rangle\langle i||j\rangle \tag{3.64}$$

$$= \sum_i \langle\phi|M_i^\dagger M_i|\psi\rangle . \tag{3.65}$$

Using the completeness relation to sum the M_i's, we have

$$(\langle\phi| \otimes \langle 0|U^\dagger)(U|\psi\rangle \otimes |0\rangle) = \langle\phi|I|\psi\rangle \tag{3.66}$$
$$= \langle\phi||\psi\rangle, \tag{3.67}$$

and so we see that the inner product is preserved, which is a defining feature of a unitary transformation.

3.7 CP-maps and positive operator-valued measurements

The unitary evolution and the projective measurements described in the last chapter are all the tools we need to describe the evolution of an isolated quantum system. CP-maps provide a convenient way of describing the nonunitary evolution of a quantum system when it interacts with other systems, for example, spontaneous emission, which is a nonunitary process owing to the nonexistence of its inverse, spontaneous absorption. Some examples of CP-maps are:

- Projections—an evolution of the form $\varrho \to \sum_i P_i \varrho P_i$ is a CP-map where the P_i's are projections. The projection measurements in the last chapter are examples of CP-maps.
- Unitary evolution is a special case of a CP-map, where only one operator is present in the sum, that is, $U\varrho U^\dagger$. (Thus CP-maps generalize both measurements and unitary operations).
- Addition of another system to ϱ is also a CP-map: $\varrho \to \varrho \otimes \sigma$;
- Let $\sum_i E_i = I$. Then $\varrho \to p_i := \mathrm{tr}(\varrho E_i)$ is a CP-map if $E_i \geq 0$, which generates a probability distribution from a density matrix.

A positive map is a map which maps density operators to density operators. Remarkably, not all positive maps are completely positive, transposition being a well-known example. The positivity of transposition follows from the fact for any state ϱ, its transpose $\varrho^T \geq 0$. However, a counterexample to the assertion of completeness is provided by the example of a singlet state consisting of two subsystems A and B. Namely, if we transpose only A (or only B), then the resulting operator is not positive (so that it is not a physical state); that is, $\varrho_{AB}^{T_A} < 0$. Confirmation of this is left as an exercise. This fact will be crucial for quantifying entanglement in Chapter 10.

Finally, I would like to discuss another frequently used concept that is in some sense derived from the notion of CP-maps. It can be loosely stated that a CP-map represents the evolution of a quantum system when we do not update our knowledge of its state on the basis of the outcome of a particular measurement. This is why we have a summation over all measurements. If, on the other hand, we know that the outcome corresponding to the operator $M_j^\dagger M_j$ occurs, then the state of the system immediately afterwards is given by $M_j \varrho M_j^\dagger / \mathrm{tr}(M_j^\dagger M_j \varrho)$. This type of measurement is the most general one and is commonly referred to as a positive operator valued measure (POVM). It is positive because operators of the form $M^\dagger M$ are always positive for any operator M, and taking the trace of it together with any density matrix generates a positive number (i.e. a probability for that particular measurement outcome). The concept of a POVM plays a significant role in quantum information theory.

3.8 The postulates of quantum mechanics revisited

With the extra mathematics that we have learned in this chapter, we are ready to restate the first, third, and fourth postulates of quantum mechanics in a more general form.

- *States.* States are represented by density operators over a Hilbert space. A density operator can be used to represent a probability distribution of quantum states. This new postulate says that the state of a quantum system is in general a probability distribution of pure quantum states.
- *Observables.* Observables are represented by Hermitian operators as before.
- *Measurements.* A measurement is a set of positive operators $E_1, \ldots, E_k$ which satisfies the completeness relation $\sum_i E_i = I$. When a measurement is made on a quantum state ϱ it has an outcome i with a probability $\text{tr}(E_i \varrho)$. These operators can always be written in the following form: $E_i = M_i^\dagger M_i$ for some operator M_i.
- *Evolution.* The state of a quantum system evolves under a completely positive trace-preserving map.

These are the most general postulates of quantum mechanics and, surprisingly, never appear in any textbooks on the subject. There are two reasons for this. First of all, some people consider the present topic too difficult. However, more importantly, more people are not even aware of them, even when they work within quantum mechanics and its applications! These axioms are, however, extremely important for us and will be exploited extensively in the remaining part of the book.

3.9 Summary

Quantum bits cannot be cloned: given an unknown quantum state $|\psi\rangle = \alpha|0\rangle + \beta|1\rangle$, it is impossible to produce two copies $|\psi\rangle|\psi\rangle$ of $|\psi\rangle$.

Two parties can use quantum bits to detect eavesdropping when they are creating a random classical string known to both parties but not to anyone else. This random string is often called a key. This shared key can then be used by the two parties to perform a cryptographic protocol to prevent anyone who intercepts their messages from being able to read the content of the messages.

The trace of a density operator is defined as

$$\text{tr}(\varrho) = \sum_{k=1}^{n} \langle k|\varrho|k\rangle , \qquad (3.68)$$

where ϱ lies in the support of the basis $\{k\}_{k=1}^n$. We have $\text{tr}(\varrho) = 1$ if ϱ is a quantum state, $\text{tr}(\varrho^2) = 1$ if and only ϱ is a pure state. The partial-trace operator of a density operator $\varrho^{AB} = \varrho^A \otimes \varrho^B$, possibly entangled over two systems A and B, is

$$\text{tr}_B(\varrho^{AB}) = \varrho^A \otimes \text{tr}(\varrho^B) , \qquad (3.69)$$

which traces out system B and leaves behind system A. If ϱ^A and ϱ^B are not entangled, then

$$\text{tr}_B(\varrho^{AB}) = \text{tr}_B(\varrho^A \varrho^B) = \varrho^A . \qquad (3.70)$$

The purification of a density operator $\varrho^A = \sum_i p_i |i^A\rangle$ is the pure state

$$|\psi^{AB}\rangle = \sum_i \sqrt{p_i}|i^A i^B\rangle \ . \tag{3.71}$$

The density operator for A can be obtained from $|\psi^{AB}\rangle$ by tracing out B. Any mixture can be considered as a pure state in an extended Hilbert space.

The Schmidt decomposition is a process by which we can write a pure state $|\psi^{AB}\rangle$ over two systems A and B as

$$|\psi^{AB}\rangle = \sum_i \alpha_i |\psi_i^A\rangle |\phi_i^B\rangle \ , \tag{3.72}$$

where the $|\psi_i^A\rangle$'s and the $|\phi_i^B\rangle$'s form bases for systems A and B, respectively. The Schmidt decomposition writes a pure state in terms of a purification, from which many properties such as the density operators of either system, can be easily obtained.

By combining projective measurements with unitary operations on an auxiliary Hilbert space, we obtain the most general form of measurement (or nonunitary evolution), which is

$$\varrho \to \sum_i M_i \varrho M_i^\dagger \ , \tag{3.73}$$

where the M_i's satisfy the completeness relation $\sum_i M_i = I$. Under such a measurement, the system ϱ collapses into the state $|i\rangle$ with probability

$$P(i) = \mathrm{tr}(\varrho M_i^\dagger M_i) \ . \tag{3.74}$$

We can use these most general measurements to replace the projective measurements of the previous chapter in the postulates of quantum mechanics.

The most general physical evolution consistent with quantum mechanics is a completely positive, trace-preserving linear map (called a CP-map for short). It will be discussed and used in much more detail in Chapters 6, 8, 9, and 10.

4
Quantum communication with entanglement

In the previous chapter, we discussed a simple example of quantum communication which did not use entanglement—the BB84 cryptography protocol. In this chapter, we discuss some of the ways in which entanglement can be used for communication. Entanglement is a purely quantum phenomenon and none of the protocols we discuss are possible classically. We shall be able to explain quite a few protocols without really understanding what entanglement is and how to quantify it. These protocols themselves will in fact show us how different entanglement is from anything we have seen in classical physics.

If Alice and Bob share an entangled state, they can use it for dense coding, in which Alice sends Bob just one qubit of an entangled pair, which delivers two bits of classical information. Alice and Bob can also use an entangled state for teleportation, in which a qubit in a unknown quantum state is teleported from Alice to Bob when Alice sends Bob two classical bits.

Entanglement swapping is an example of a tripartite communication protocol. If Alice and Bob share entanglement, and Bob and Charlie share entanglement, entanglement swapping can be used to entangle Alice and Charlie.

Each of these communication protocols requires some classical communication and so does not break the laws of relativity, which say that nothing can travel faster than the speed of light. We show that even if Alice and Bob are entangled, Alice cannot communicate with Bob without sending him classical or quantum bits, which prevents Alice and Bob from breaking the laws of relativity.

4.1 Pure state entanglement and Pauli matrices

At the end of Chapter 1, we briefly discussed entanglement. A quantum state $|\psi_{AB}\rangle$ shared by two parties Alice and Bob is entangled if it cannot be expressed as a composition of two states $|\psi_A\rangle \otimes |\psi_B\rangle$. The Bell states are four orthogonal entangled states which we shall use in the protocols in this chapter. The four Bell states are

$$|\Phi_{AB}^{\pm}\rangle = \frac{1}{\sqrt{2}}(|0^A 0^B\rangle \pm |1^A 1^B\rangle), \tag{4.1}$$

$$|\Psi_{AB}^{\pm}\rangle = \frac{1}{\sqrt{2}}(|0^A 1^B\rangle \pm |1^A 0^B\rangle). \tag{4.2}$$

The Pauli operators (also called Pauli matrices) are

$$\sigma_1 = |1\rangle\langle 0| + |0\rangle\langle 1|, \tag{4.3}$$
$$\sigma_2 = i|1\rangle\langle 0| - i|0\rangle\langle 1|, \tag{4.4}$$
$$\sigma_3 = |0\rangle\langle 0| - |1\rangle\langle 1|. \tag{4.5}$$

Alice (or Bob) can apply one or more of the Pauli operators locally to change between any of the Bell states:

$$(\sigma_1 \otimes I)|\Phi^{\pm}_{AB}\rangle = |\Psi^{\pm}_{AB}\rangle, \tag{4.6}$$
$$(\sigma_1 \otimes I)|\Psi^{\pm}_{AB}\rangle = |\Phi^{\pm}_{AB}\rangle, \tag{4.7}$$
$$(\sigma_3 \otimes I)|\Phi^{\pm}_{AB}\rangle = |\Phi^{\mp}_{AB}\rangle, \tag{4.8}$$
$$(\sigma_3 \otimes I)|\Psi^{\pm}_{AB}\rangle = |\Psi^{\mp}_{AB}\rangle. \tag{4.9}$$

The effect of σ_2 is basically the same as that of the product of σ_1 and σ_3.

If a state $|\psi^{AB}\rangle$ is entangled, then tracing out one of the two systems leads to a mixed state. If $|\psi_{AB}\rangle = |\psi_A\rangle \otimes |\psi_B\rangle$ (i.e. it is not entangled), then tracing out part A or part B of the space leaves $|\psi_B\rangle$ or $|\psi_A\rangle$, respectively. Recalling that $\text{tr}(\varrho^2) = 1$ if and only if ϱ is a pure state, we have a simple formula for testing whether a state is entangled or not. The state $|\psi_{AB}\rangle$ is entangled if and only if

$$\text{tr}(\text{tr}_A(|\psi^{AB}\rangle\langle\psi^{AB}|)^2) < 1. \tag{4.10}$$

All of the Bell states are entangled. For example

$$\text{tr}(\text{tr}_B(|\Phi^{\pm}_{AB}\rangle\langle\Phi^{\pm}_{AB}|)^2) = \text{tr}(\text{tr}_B(|\Phi^{\pm}_{AB}\rangle\langle\Phi^{\pm}_{AB}|)^2) \tag{4.11}$$
$$= \text{tr}\left(\frac{1}{2}|0\rangle\langle 0| + \frac{1}{2}|1\rangle\langle 1|\right) \tag{4.12}$$
$$= \frac{1}{4} + \frac{1}{4} \tag{4.13}$$
$$< 1. \tag{4.14}$$

However, this simple way of detecting entanglement doesn't work for mixed states.

4.2 Dense coding

Dense coding is a simple yet nontrivial example of how entanglement can be used in quantum communication. Dense coding works by Alice using the Pauli operators to change the shared state of Alice and Bob between the four Bell states. Using dense coding, Alice can send Bob two bits of information by sending him one qubit, which is her half of an entangled Bell state. By performing a communication that cannot be performed classically (i.e. encoding two bits of classical information into one qubit), dense coding provides a convincing example that quantum information differs from any sort of classical information.

Initially Alice and Bob share the Bell state

$$|\Phi^+\rangle = \frac{1}{\sqrt{2}}(|0^A 0^B\rangle + |1^A 1^B\rangle). \tag{4.15}$$

Alice also has two bits x and y of classical information which she would like to send to Bob. Alice and Bob have agreed in advance some unitary operations that Alice will perform depending on the values of x and y. Alice reads the first bit x. If $x = 0$, she does nothing. If $x = 1$, she performs a swap operation σ_1 on her qubit which transforms $|\Phi^+\rangle$ to

$$(\sigma_1 \otimes I)|\Phi^+\rangle = |\Psi^+\rangle . \tag{4.16}$$

Alice now reads the second bit y. If $y = 0$, she does nothing. If $y = 1$, she performs a phase shift σ_3 on her qubit. The phase shift changes $|\Phi^+\rangle$ and $|\Psi^+\rangle$ to

$$(\sigma_3 \otimes I)|\Phi^+\rangle = |\Phi^-\rangle \tag{4.17}$$

and

$$(\sigma_3 \otimes I)|\Psi^+\rangle = |\Psi^-\rangle . \tag{4.18}$$

Depending on the values of x and y, Alice and Bob now share one of the four Bell states. The state that either Alice or Bob sees is a maximally mixed state

$$\mathrm{tr}_A |\Phi^\pm\rangle\langle\Phi^\pm| = \mathrm{tr}_A |\Psi^\pm\rangle\langle\Psi^\pm| = \mathrm{tr}_B |\Phi^\pm\rangle\langle\Phi^\pm| = \mathrm{tr}_B |\Psi^\pm\rangle\langle\Psi^\pm| = \frac{1}{2}|0\rangle\langle 0| + \frac{1}{2}|1\rangle\langle 1| . \tag{4.19}$$

Alice and Bob cannot deduce from measurements on their own systems which Bell state they share. However, Alice can send Bob her qubit, in which case Bob has one of the four orthogonal Bell states, which he can measure and can then deduce the values of x and y.

4.3 Teleportation

Teleportation is a process by which Alice can send Bob one qubit in an unknown state $|\psi\rangle$ by sending Bob two classical bits if Alice and Bob initially share an entangled Bell state. When classical bits are sent over a classical channel, it is possible for Alice to retain a copy. However, the no-cloning theorem says that it is impossible for Alice to copy the unknown state $|\psi\rangle$. When she sends $|\psi\rangle$ to Bob, she retains no information about the state of $|\psi\rangle$—it is as if the state $|\psi\rangle$ moves from Alice to Bob—hence the name, teleportation.

Initially, Alice and Bob share an entangled Bell state

$$|\Phi^+_{AB}\rangle = \frac{(|0_A\rangle|0_B\rangle + |1_A\rangle|1_B\rangle)}{\sqrt{2}} . \tag{4.20}$$

Alice also has a qubit in an unknown state $|\psi\rangle = \alpha|0\rangle + \beta|1\rangle$, and she wants to teleport it to Bob as in Fig. 4.1. The state has to be unknown to her, because otherwise she can just phone Bob up and tell him all the details of the state and he can then recreate it with a particle that he possesses. Given that Alice does not know the state, she cannot measure it to obtain all the information necessary to specify it.

Initially, the state of all three qubits is

$$|\psi_A\rangle|\Phi^+_{AB}\rangle = (\alpha|0_A\rangle + \beta|1_A\rangle)\frac{1}{\sqrt{2}}(|0_A 0_B\rangle + |1_A 1_B\rangle) . \tag{4.21}$$

We can expand and rewrite this as

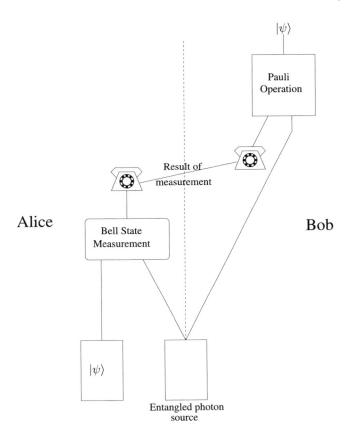

Fig. 4.1 An unknown quantum state $|\psi\rangle$ is teleported from Alice to Bob. Initially, Alice has the unknown state $|\psi\rangle$ and one half of an entangled pair of photons. Bob has the other photon. Alice performs a measurement on her system and sends the result to Bob. Bob uses the result to perform a unitary operation on his photon, which changes its state to $|\psi\rangle$. Alice cannot determine the state of $|\psi\rangle$ and, because of the no-cloning theorem, no record of $|\psi\rangle$ is kept by Alice. So $|\psi\rangle$ has been teleported from Alice to Bob.

$$\begin{aligned}|\psi_A\rangle|\Phi^+_{AB}\rangle &= \frac{1}{\sqrt{2}}(\alpha|0_A 0_A 0_B\rangle + \alpha|0_A 1_A 1_B\rangle + \beta|1_A 0_A 0_B\rangle + \beta|1_A 1_A 1_B\rangle) \\ &= \frac{1}{2}[|\Phi^+_{AA}\rangle(\alpha|0_B\rangle + \beta|1_B\rangle) + |\Phi^-_{AA}\rangle(\alpha|0_B\rangle - \beta|1_B\rangle) \\ &+ |\Psi^+_{AA}\rangle(\alpha|1_B\rangle + \beta|0_B\rangle) + |\Psi^-_{AA}\rangle(\alpha|1_B\rangle - \beta|0_B\rangle)]\ . \end{aligned} \quad (4.22)$$

Alice's system (two qubits) is now written in terms of the four Bell states. The state of Bob's qubit in all four cases "looks very much like" the original qubit that Alice has to teleport to Bob. Alice now measures her part of the system in the Bell basis. She randomly, obtains, one of the Bell states and uses two bits to send Bob the result of the measurement. Bob now knows which of the four states $\alpha|0\rangle \pm \beta|1\rangle$, $\alpha|1\rangle \pm \beta|0\rangle$ he has and can apply a unitary operation to his system to obtain the original $|\psi\rangle$. Table

Alice's measurement	Bob's state	Bob's operation
$\|\Phi^+\rangle$	$\alpha\|0\rangle + \beta\|1\rangle$	I
$\|\Phi^-\rangle$	$\alpha\|0\rangle - \beta\|1\rangle$	σ_3
$\|\Psi^+\rangle$	$\alpha\|1\rangle + \beta\|0\rangle$	σ_1
$\|\Psi^-\rangle$	$\alpha\|1\rangle - \beta\|0\rangle$	$\sigma_1\sigma_3$

Table 4.1

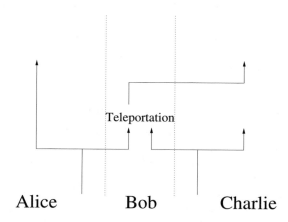

Fig. 4.2 Entanglement swapping. Initially, entangled pairs are shared between Alice and Bob, and between Bob and Charlie. There is no entanglement between Alice and Charlie. However, Bob can use his entanglement with Charlie to teleport his entanglement with Alice to Charlie. Thus Alice and Charlie become entangled even though they have never directly communicated.

4.1 shows which operation Bob performs.

4.4 Entanglement swapping

Can systems that have never interacted become entangled? The answer is, very surprisingly, "yes" and entanglement swapping is the method.[1] Accordingly, if A and D have never interacted, it is enough that A is entangled with B and D with C, and we then measure C and D by projecting them into an entangled state. The result will be that A and D will also become entangled.

"Entanglement swapping" is closely related to quantum teleportation. Whereas quantum teleportation enables the state of a system (e.g. a particle or collection of

---

[1] How surprising this really is was first beautifully illustrated by a friend of mine, Sougato Bose, using the following analogy. Imagine that Alice and Bob are married (entangled) and so are Charles and Diana (the names are chosen so that the initials are A, B, C, D, and any similarities to real-life characters are purely accidental). Imagine, furthermore, that the two couples initially do not know each other. Now suppose that Alice and Charles meet at a party, start to talk to each other and—after some time—start to become entangled with each other. It would be quite surprising in the everyday world if their respective partners, Bob and Diana (who have never met each other), were also to start to become entangled as a consequence. And yet this is exactly what entanglement swapping is all about—a kind of quantum mechanical Woody Allen type of "partner-swapping" story.

particles) to be teleported to an independent physical system via classical communication channels and a shared entanglement resource, the purpose of entanglement swapping is to induce entanglement between systems that hitherto have shared no entanglement. An entanglement resource is required for entanglement swapping to occur; indeed, the nomenclature "entanglement swapping" describes the transfer of entanglement from a priori entangled systems to a priori separable systems.

The relationship between teleportation and entanglement is shown in Fig. 4.2. If Alice and Bob share a Bell state such as $|\Phi^-_{AB}\rangle$ and Bob and Charlie share a Bell state such as $|\Phi^+_{BC}\rangle$, then Bob and Charlie can use $|\Phi^+_{BC}\rangle$ to teleport Bob's part of $|\Phi^-_{AB}\rangle$ to Charlie (this uses classical communication). At the end of the protocol, Alice and Charlie share the Bell state $|\Phi^-_{AB}\rangle$. The entanglement between Bob and Charlie is destroyed in the teleportation process.

4.5 No instantaneous transfer of information

Faster-than-light communication would break the laws of relativity. In the above protocols, classical bits were sent between Alice and Bob, preventing them from signaling to one another faster than the speed of light. Is there any quantum mechanical way of performing faster-than-light communication?

Suppose Alice and Bob share a mixed state $\varrho_{AB} = \sum_{i,j} p_{ij} |\psi_i^A\rangle \otimes |\phi_j^B\rangle$ and Alice performs a general operation on her system specified by the measurement operators M_i, where $\sum_i M_i = I$. Is it possible for Alice to influence Bob's system to send him information without interacting with him? The state of Bob's system after Alice's operation is

$$\varrho'_B = \mathrm{tr}_A \left(\sum_k (M_k \otimes I) \varrho_{AB} (M_k^\dagger \otimes I) \right) \quad (4.23)$$

$$= \mathrm{tr}_A \left(\sum_{i,j,k} p_{ij} M_k |\psi_i^A\rangle M_k^\dagger \otimes |\phi_j^B\rangle \right) \quad (4.24)$$

$$= \mathrm{tr} \left(p_{ij} M_k |\psi_i^A\rangle M_k^\dagger \right) \quad (4.25)$$

$$= \sum_i \mathrm{tr}_A \left((M_i \otimes I) \varrho_{AB} (M_i^\dagger \otimes I) \right) \quad (4.26)$$

$$= \sum_i \mathrm{tr}_A \left((M_i M_i^\dagger \otimes I) \varrho_{AB} \right) \quad (4.27)$$

$$= \mathrm{tr}_A(\varrho_{AB}) \quad (4.28)$$

$$= \varrho_B . \quad (4.29)$$

So whatever Alice does, she does not change the state of Bob's system, and no faster-than-light communication is possible (strictly speaking, no instantaneous communication is possible—the speed of light does not enter this argument in any way). Thus quantum mechanics is a local theory. There is no way of instantaneously influencing something over there by acting over here.

4.6 The extended–Hilbert–space view

It is now clear that a typical quantum information-processing protocol involves entangled quantum systems, quantum measurements on them, and—very importantly—classical communication. We emphasize that for protocols such as teleportation and entanglement swapping, classical communication is absolutely crucial. Without it, the final state of the system would not be known (since the classical communication conveys the outcome of a measurement) and would in fact be a maximally mixed state. As such, it would not be useful for any quantum information processing.

But what is classical communication in the first place? We have been arguing in this book that the world is—as far as we can tell so far—quantum mechanical and that classical physics is only an approximation to reality. Classical communication is therefore a very special type of communication, where we do not use the effects of superpositions.[2]

I shall now present a more coherent view of teleportation where there is no division between quantum and classical information and where everything is quantum mechanical—as it should be. Suppose, therefore, that Alice shares with Bob a Bell state and that she, in addition, has an extra qubit in an unknown state that needs to be teleported to Bob. Now, Alice makes a measurement in the Bell basis on her two qubits. Suppose that this measurement is actually written into another (four-level) quantum system. This four-level quantum system is then sent to Bob. Conditionally on the state of the four-level system, Bob performs one (or none) of the Pauli operations. And this results in teleportation. There are no measurements involved here, only conditional unitary operations (which themselves are unitary by definition).

The reason why a qubit going from Alice to Bob in a mixture of states numbered from one to four can be considered a classical communication is that we never use a superposition of these states; we only use the numbers themselves. So, effectively, this qubit need not be a qubit — it can just be an ordinary classical bit![3]

4.7 Summary

Entanglement is a purely quantum resource which is not available in classical information theory. Entanglement can be used to communicate in ways which are not possible classically.

The Bell states are four orthogonal, fully entangled pairs of qubits

$$|\Phi^{\pm}_{AB}\rangle = \frac{1}{\sqrt{2}}(|0^A 0^B\rangle \pm |1^A 1^B\rangle), \quad (4.30)$$

$$|\Psi^{\pm}_{AB}\rangle = \frac{1}{\sqrt{2}}(|0^A 1^B\rangle \pm |1^A 0^B\rangle). \quad (4.31)$$

The Pauli matrices are

---

[2]This statement is not well defined. What is a superposition in one basis is not a superposition in another. However, in reality, one basis may be preferred to another for physical reasons.

[3]It could, of course, happen that Hilbert space cannot be extended. This is why the belief in a Hilbert space that is always extendible has been called "the Church of Higher Hilbert Space" by Ben Schumacher, the discoverer of the qubit.

$$\sigma_1 = |1\rangle\langle 0| + |0\rangle\langle 1|, \qquad (4.32)$$
$$\sigma_2 = i|1\rangle\langle 0| - i|0\rangle\langle 1|, \qquad (4.33)$$
$$\sigma_3 = |0\rangle\langle 0| - |1\rangle\langle 1|. \qquad (4.34)$$

Dense coding is a process by which one party can send another two classical bits of information by sending one quantum bit if, prior to the communication, the two parties shared an (entangled) Bell state. The Bell state is destroyed by this process.

Teleportation is a means by which one party can teleport an unknown quantum bit if the two parties share a Bell state, by sending two bits of classical information. Teleportation is so called because the sender destroys their copy of the unknown quantum bit and is left with no knowledge of its value. (Remember that quantum bits cannot be cloned).

Entanglement provides a protocol based on teleportation. If systems A and B are entangled and systems B and C are entangled, then the entanglement between systems B and C can be used to teleport a qubit from system B to system C. If this qubit of system B is entangled with system A, then after the teleportation, systems A and C are entangled, despite there not having been any communication between systems A and C.

The above protocols rely on classical communication, which cannot be performed faster than the speed of light. Communication faster than the speed of light would violate the laws of relativity. It can be proved that there are no quantum operations, even using entanglement, which allow information to be carried faster than the speed of light.

We can consider classical information theory as a subset of quantum information theory where we are restricted to orthogonal states. In this view, there is no division between the classical and quantum worlds. When we talk about classical communication, we mean quantum communication which does not use the superposition principle.

5
Quantum information I

We have now worked a great deal with quantum bits, which make up what we call quantum information. How can quantum information be characterized? We usually think of information as being classical — in a definite state rather than in a superposition of states. It seems rather strange to consider information in superpositions.[1] Some people would, on the basis of this argument, conclude that quantum information can never exist and we can only have access to classical information.[2] It turns out, however, that quantum information can be quantified in the same way as classical information using Shannon's prescription. It can be shown that there is a unique measure (up to a constant additive or multiplicative term) of quantum information such that

- S is purely a function of the probabilities of outcomes of measurements made on a quantum system (i.e. a function of a density operator);
- S is a continuous function of probability;
- S is additive; that is, if ϱ_1 and ϱ_2 are uncorrelated systems, then

$$S(\varrho_1 \otimes \varrho_2) = S(\varrho_1) + S(\varrho_2) \,. \tag{5.1}$$

The unique measure S is called the von Neumann entropy.[3] The von Neumann entropy of a density operator ϱ, written in diagonal form (using a spectral decomposition) as

$$\varrho = \sum_i p_i |i\rangle\langle i| \,, \tag{5.2}$$

is

$$S(\varrho) = -\text{tr}(\varrho \log \varrho) \,, \tag{5.3}$$

which can be reformulated as

$$S(\varrho) = -\sum_i p_i \log(p_i) \,. \tag{5.4}$$

This looks rather like the formula for the Shannon entropy. If X is a random variable which takes values i with probability p_i, then the von Neumann and Shannon entropies

---

[1]The usual argument as to why this is strange is because we never see superpositions of states of objects in everyday world, which is where we receive information. The statement that we do not "see" superpositions is, of course, rather vague and dubious.

[2]This would, somewhat discouragingly, make my position at Leeds University imaginary, since I am a professor of quantum information!

[3]This argument will be used again when quantifying the entanglement of pure states.

are the same, that is, $H(X) = S(\varrho)$. This is the case because we can think of a density operator in its diagonal basis as a classical probability distribution. The proof of the uniqueness of the von Neumann entropy goes very much along the lines of Shannon's proof, and we leave it to the reader to fill in the gaps.

Analogously to Shannon's classical entropy, we can show that the von Neumann entropy is the expected rate of compression of a string of quantum states, that is, density operators. Before doing this, we shall explain the first measure that, loosely speaking, quantifies the nonclassicality of quantum information—the fidelity. This quantity measures how close to each other two quantum states are.

5.1 Fidelity

Let us start our discussion with pure states, as we have already met fidelity for pure states, but in a slightly different guise. If $|\psi\rangle$ and $|\phi\rangle$ are two pure states then the probability of measuring $|\psi\rangle$ to obtain $|\phi\rangle$ or vice versa is

$$F(|\psi\rangle, |\phi\rangle) = |\langle\phi|\psi\rangle|^2 . \qquad (5.5)$$

We call F the **fidelity** between $|\psi\rangle$ and $|\phi\rangle$. It measures the probability of confusing the two states if we are allowed to make only one measurement on one system, prepared in one of the states. The fidelity is minimal (equal to 0) when $|\psi\rangle$ and $|\phi\rangle$ are orthogonal and can be fully distinguished—that is, we can never confuse the states—and maximal (equal to 1) when $|\psi\rangle = |\phi\rangle$ and they are completely indistinguishable. The fact that there is a continuum of values of the fidelity between zero and one shows us that quantum states of maximal knowledge (i.e. pure states) are different from classical states of maximal knowledge. Classically, our system could, for example, move with velocity of $5ms^{-1}$ or $6ms^{-1}$. But if the system moves at $5ms^{-1}$, then it is not moving at $6ms^{-1}$ (definitely). The distinction between 5 and 6 is in this case absolute and, classically (at least in principle, when we have maximal information) these states cannot be confused in any way.

We can also define the fidelity between a pure state and a mixed state as

$$F(|\psi\rangle, \varrho) = \langle\psi|\varrho|\psi\rangle , \qquad (5.6)$$

which is, likewise, the probability of measuring ϱ as (or confusing with) $|\psi\rangle$. This gives us the probability that if our system is prepared in the state ϱ, we shall measure it[4] in the state $|\psi\rangle$.

If $\varrho = \sum_i p_i |i\rangle\langle i|$ is a mixture, then the fidelity between ϱ and another mixture σ is average fidelity between $|i\rangle$ and σ, given by

$$F(\sigma, \varrho) = \sum_i p_i F(|i\rangle, \sigma) = \sum_i p_i \langle i|\sigma|i\rangle . \qquad (5.7)$$

If $\sigma = \sum_j q_j |q_j\rangle\langle q_j|$, then this fidelity can be reformulated as

$$F(\sigma, \varrho) = \sum_{ij} p_i q_j |\langle i|q_j\rangle|^2 = \text{tr}(\varrho\sigma) . \qquad (5.8)$$

---

[4] This is true from one of the axioms of quantum mechanics, related to the expected value of a Hermitian operator in a given quantum state.

54   *Quantum information I*

If $\varrho = \sigma$ then $F(\sigma, \varrho) = \text{tr}(\varrho^2)$. Because of the fact that this is not always equal to unity, some people like to define the fidelity differently as $\text{tr}(\sqrt{\varrho\sigma})$, which still gives the same value as before for pure states. ϱ and σ can be distinguished when they have orthogonal supports (they are in orthogonal vector spaces), in which case $F(\varrho, \sigma) = 0$, no matter which definition of the fidelity we choose. The exact definition of the fidelity for mixed states will not concern us here, as all different (reasonable) definitions will have the same properties when applied to the things that matter to us in this book.

5.2   Helstrom's discrimination

Helstrom's discrimination is an excellent instance of a protocol that illustrates the concept of distinguishability of quantum states and shows exactly how quantum information is different from classical information.

The problem in question is the following. Suppose that someone prepares one of the two (nonorthogonal in general) pure states $|\psi_1\rangle$ and $|\psi_2\rangle$, with equal probabilities of $1/2$ for the two states. What is the best strategy for us to follow to discriminate the two states; that is, what is the highest probability of being able to discriminate between the two states? Note that this is not quite the same as the fidelity between the two states.

Let us first argue about the best strategy in a physically intuitive way and then prove the result more formally. The measurement that we should set up should have two orthogonal outcomes $|m_1\rangle$ and $|m_2\rangle$, each corresponding to one of the two states. If we happen to project onto $|m_1\rangle$, then we guess that the prepared state was $|\psi_1\rangle$, while if we project onto $|m_2\rangle$, we guess that the prepared state was $|\psi_2\rangle$. The probability of success is then

$$p_S = \frac{1}{2}(|\langle\psi_1|m_1\rangle|^2 + |\langle\psi_2|m_2\rangle|^2) . \tag{5.9}$$

The overall factor of one-half in front is there because the two states have this equal probability of preparation. Now, it is not so surprising that the maximum of the above quantity is achieved when the two quantities in the brackets are equal. Indeed, the situation is symmetric with respect to exchanging $|\psi_1\rangle$ and $|\psi_2\rangle$ and this must be reflected in the probability of success. The highest that can be achieved is

$$p_S = \frac{1}{2}(1 + \sqrt{1 - |\langle\psi_1|\psi_2\rangle|^2}) . \tag{5.10}$$

If the two states are orthogonal, the probability for discrimination is unity (as it should be) and when the states are one and the same, we cannot succeed at discriminating between them at all (as also expected).

Let us now prove this formally. The probability of success, when we have two measurement operators $M_1 = |m_1\rangle\langle m_1|$ and $M_2 = |m_2\rangle\langle m_2|$ (such that $M_1 + M_2 = I$ for completeness), is

$$\begin{align}
p_S &= \frac{1}{2}(\langle\psi_1|M_1|\psi_1\rangle + \langle\psi_2|M_2|\psi_2\rangle) \tag{5.11}\\
&= \frac{1}{2}(1 + \langle\psi_2|M_2|\psi_2\rangle - \langle\psi_1|M_2|\psi_1\rangle) \tag{5.12}\\
&= \frac{1}{2}(1 + \text{tr}\, M_2(|\psi_2\rangle\langle\psi_2| - |\psi_1\rangle\langle\psi_1|)) . \tag{5.13}
\end{align}$$

The trick therefore is to maximize the quantity tr $M\Gamma$, where $\Gamma = |\psi_2\rangle\langle\psi_2| - |\psi_1\rangle\langle\psi_1|$. This we do by finding the eigenvectors of Γ, and M_2 should then be the projection onto the eigenvector with a positive eigenvalue (the other eigenvector—there will only be two—will have a negative eigenvalue). The eigenvalues of Γ are $\pm\sqrt{1 - |\langle\psi_1|\psi_2\rangle|^2}$. We obtain the final probability in terms of the overlap of the two states.

I would like, finally, to point out that this problem can also be solved in general if the two given states are mixed and appear with different probabilities. This generalization, however, will not concern us here, suffice it to say that it can be solved by diagonalizing the difference between the two density matrices representing the mixtures.[5]

5.3 Quantum data compression

The optimal method of communication through a noiseless channel using *pure* states is equivalent to data compression. In Chapter 1 we saw that the limit on classical data compression is given by the entropy of the probability distribution of the data. We would thus guess that the limit on quantum data compression is given by the von Neumann entropy of the set of states being compressed. This, in fact, turns out to be a correct guess as first proven by Schumacher (1995).[6] So, Alice now encodes the letters of her classical message into pure quantum states and sends these to Bob. For example, if $0 \to |\psi_0\rangle$ and $1 \to |\psi_1\rangle$, then Alice's message 001 will be sent to Bob as the sequence of pure quantum states $|\psi_0\rangle|\psi_0\rangle|\psi_1\rangle$.

The exact problem can be phrased in the following equivalent fashion: suppose a quantum source randomly prepares different qubit states $|\psi_i\rangle$ with corresponding probabilities p_i. A random sequence of n such states is produced. By how much can this be compressed, that is, how many qubits do we really need to encode the original sequence (in the limit of large n)? First of all, the total density matrix is

$$\varrho = \sum_i p_i |\psi_i\rangle\langle\psi_i| .$$

Now, this matrix can be diagonalized.

$$\varrho = \sum_i q_i |\phi_i\rangle\langle\phi_i| ,$$

where q_i and $|\phi_i\rangle$ are eigenvectors and eigenvalues. This decomposition is, of course, indistinguishable from the original one (or any other decomposition for that matter[7]). Thus we can think about compression in this new basis, which is easier as it behaves completely classically (since $\langle\phi_i|\phi_j\rangle = \delta_{ij}$). We can therefore apply the results from the previous section to classical typical sequences and conclude that the limit on compression is $n(-\sum_i q_i \log q_i)$, that is, n qubits can be encoded into $nS(\varrho)$ qubits.

---

[5]The thesis of Chris Fuchs (1996) is an excellent and detailed introduction to all these distinguishability issues.

[6]I believe that he proved this in 1993, but the paper was published in Phys. Rev. A only in 1995.

[7]This fact that a given density matrix has infinitely many indistinguishable decompositions is very important in quantum mechanics, and the fact that we cannot obtain instantaneous communication using entangled quantum systems rests on it (among many other facts).

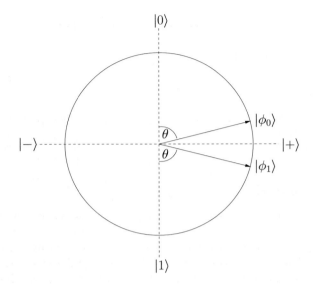

Fig. 5.1 Two nonorthogonal states used to encode a message. The overlap between them is $\sin\theta$. The larger the overlap, the more the total message can be compressed. However, they can be compressed since the larger the overlap, the less information they carry.

No matter how the states are generated, as long as the total state is described by the same density matrix ϱ, its compression limit is its von Neumann entropy. This protocol and result will be very important when we discuss measures of entanglement.
Example. Suppose that Alice encodes her bit into states $|\phi_0\rangle = \cos(\theta/2)|0\rangle + \sin(\theta/2)|1\rangle$ and $|\phi_1\rangle = \sin(\theta/2)|0\rangle + \cos(\theta/2)|1\rangle$, where $p_0 = p_1 = 1/2$ (see Fig. 5.1). Classically, it is not possible to compress a source that generates 0 and 1 with equal probability. In quantum mechanics, we can. However, since $|\phi_0\rangle$ and $|\phi_1\rangle$ are nonorthogonal, we cannot distinguish them with complete accuracy — the amount of extra compression gained can be seen as the information that we lose about the original message 001. In our example, the overlap between the two states is $\langle\phi_0|\phi_1\rangle = \sin\theta$, and they are orthogonal only when $\theta = \pi$, in which case no compression is possible. Otherwise, the compression rate increases with the overlap between the states.

Suppose Alice's messages are only three qubits long. Then there are eight different possibilities, $|\phi_0\phi_0\phi_0\rangle, \ldots, |\phi_1\phi_1\phi_1\rangle$, which are all equally likely with a probability of $1/8$. In general, these states will lie, with a high probability, within a subspace of the eight-dimensional Hilbert space. Let us call this likely subspace the "typical" subspace. Its orthogonal complement will be unlikely and hence is called the "atypical" subspace. In order to find the typical and atypical subspaces, we can consider the density operator

$$\varrho = \frac{1}{2}(|\phi_0\rangle\langle\phi_0| + |\phi_1\rangle\langle\phi_1|) \,. \tag{5.14}$$

We look at it in its diagonal form (the reader can check this by expanding the two expressions for ϱ),

$$\varrho = \frac{1+\sin\theta}{2}|+\rangle\langle+| + \frac{1-\sin\theta}{2}|-\rangle\langle-|, \quad (5.15)$$

where $|\pm\rangle = (|0\rangle \pm |1\rangle)/\sqrt{2}$. Now we look at the probabilities for each of the eight messages to lie in the new orthogonal basis $|+++\rangle, \ldots, |---\rangle$ of the Hilbert space of three qubits:

$$|\langle+++|\psi^{\otimes 3}\rangle|^2 = (\cos(\theta/2) + \sin(\theta/2))^6, \quad (5.16)$$
$$|\langle++-|\psi^{\otimes 3}\rangle|^2 = (\cos(\theta/2) + \sin(\theta/2))^4$$
$$* \; (\cos(\theta/2) - \sin(\theta/2))^2, \quad (5.17)$$
$$|\langle+--|\psi^{\otimes 3}\rangle|^2 = (\cos(\theta/2) + \sin(\theta/2))^2$$
$$* \; (\cos(\theta/2) - \sin(\theta/2))^4, \quad (5.18)$$
$$|\langle---|\psi^{\otimes 3}\rangle|^2 = (\cos(\theta/2) - \sin(\theta/2))^6, \quad (5.19)$$

where $|\psi^{\otimes 3}\rangle$ represents any three-qubit sequence of $|\psi_0\rangle$ and $|\psi_1\rangle$. In addition, all the probabilities for $|++-\rangle, |+-+\rangle$, and $|-++\rangle$ are equal and so are the probabilities for $|+--\rangle, |-+-\rangle$, and $|--+\rangle$. Thus the above equation contains 64 probabilities in total.

Suppose now that $\cos(\theta/2) \approx 1/\sqrt{2}$ and $\sin(\theta/2) \approx 1/\sqrt{2}$. In this case, we see that the states containing two or more $|+\rangle$'s become much more likely. This means that the message states are much more likely to be in the particular subspace spanned by the strings containing two or more $|+\rangle$'s. We can therefore compress the data as follows. First, the source generates three qubits in some state. Then we project this message onto the typical subspace using the projection operator

$$P = |+++\rangle\langle+++| + |++-\rangle\langle++-| + |+-+\rangle\langle+-+| + |-++\rangle\langle-++|. \quad (5.20)$$

If we are successful, then this will lie in the four-dimensional typical subspace, for which we need only two qubits rather than three.

We return now to considering data compression in general and proving more formally that is can be done. The first step in quantum data compression is to express the mixture to be compressed as a diagonalized density operator

$$\varrho = \sum_i p_i |i\rangle\langle i|. \quad (5.21)$$

The ϵ-typical and ϵ-atypical subspaces $\varrho^{\otimes n}$ are defined as the subspaces spanned by the ϵ-typical and ϵ-atypical sequences of X^n, where X is an analogous classical distribution with $P(X = i) = p_i$. By the analysis in Chapter 1, for any $\delta, \epsilon > 0$ there exists a sufficiently large n such that the ϵ-typical subspace has a dimension of at most $2^{n(H(X)+\delta)} = 2^{n(S(\varrho)+\delta)}$. To compress $\varrho^{\otimes n}$, so that the number of qubits the compressed state occupies is (approximately) equal to its von Neumann entropy, all we have to do is project onto this ϵ-typical subspace. The ϵ-typical subspace, which has a dimension of at most $2^{n(S(\varrho)+\delta)}$, can then be encoded by a unitary transformation into $n(S(\varrho)+\delta)$ qubits. Upon decompression, the state obtained is $\varrho^{\otimes n}$ projected onto its typical subspace.

58 *Quantum information I*

Let $\varrho_{n,\epsilon\text{-typ}}$ and $\varrho_{n,\epsilon\text{-atyp}}$ represent $\varrho^{\otimes n}$ projected onto the ϵ-typical and ϵ-atypical subspaces of $\varrho^{\otimes n}$ respectively. Again, by the analysis in Chapter 1, $\varrho^{\otimes n}$ can be written as

$$\varrho^{\otimes n} = (1-h)\varrho_{n,\epsilon\text{-typ}} + h\varrho_{n,\epsilon\text{-atyp}} \qquad (5.22)$$

where $h < \epsilon$. The probability that this projection succeeds is therefore at least $1 - \epsilon$.

Classical strings are mutually orthogonal and always have a fidelity of 0 and so a measure of the fidelity is irrelevant in the classical data compression described in Chapter 1. In the case of quantum compression, we check that the fidelity between the original and the projected state is arbitrarily high, so that upon decompression, the state obtained is arbitrarily close to the original state. The difference between fidelity and error is that the error is the probability that the decompressed state is completely different from the original state and the fidelity is the probability of being able to distinguish the decompressed state from the original state. The fidelity between the original and the projected state is

$$\begin{aligned} F(\varrho^{\otimes n}, \varrho_{n,\epsilon\text{-typ}}) &= \operatorname{tr}(((1-h)\varrho_{n,\epsilon\text{-typ}} + h\varrho_{n,\epsilon\text{-atyp}}), \varrho_{n,\epsilon\text{-typ}}) & (5.23) \\ &= 1 - h\,. & (5.24) \\ &\geq 1 - \epsilon & (5.25) \end{aligned}$$

Hence, for any $\delta > 0$, there exists a sufficiently large n such that a mixture represented by a density operator $\varrho^{\otimes n}$ can be compressed to $n(S(\varrho) + \delta)$ qubits with an arbitrarily small probability of error and an arbitrarily small probability of being able to distinguish between the original and final states.

Finally, it is important to note that even when we have a noiseless quantum channel its capacity will be reduced compared with its classical equivalent because of "quantum noise" stemming from the uncertainty principle. The very fact that different quantum states are not fully distinguishable leads to a reduction of capacity, which is why the von Neumann entropy is smaller than the corresponding Shannon entropy.

5.4 Entropy of observation

In addition to the von Neumann entropy, there is another entropy in quantum mechanics, which is due to the measurement of a particular observable (for an excellent review of quantum entropies, see Wehrl (1978)). Suppose that the state of the system is ϱ, and that we perform a measurement of the observable $A = \sum_i a_i P_i$, where the a_i's are the eigenvalues that label the outcomes (not probabilities) and the P_i's are projectors onto eigenvectors of the observable. The probability of obtaining a_j is

$$p_j = \operatorname{tr} \varrho P_j\,. \qquad (5.26)$$

The corresponding Shannon entropy of this is

$$S_\varrho(A) = -\sum_j (\operatorname{tr} \varrho P_j) \log(\operatorname{tr} \varrho P_j)\,. \qquad (5.27)$$

To see the effect of a projection on the entropy of the system, consider compressing ϱ after making a set of projective measurements. After the projection, ϱ can diagonalized as follows:

$$\varrho' = \sum_j P_j \varrho P_j \tag{5.28}$$

$$= \sum_j \mathrm{tr}(P_j \varrho) \varrho_j \tag{5.29}$$

$$= \sum_{i,j} p_j |\psi_j\rangle\langle\psi_j| \tag{5.30}$$

where the $|\psi_j\rangle$'s are orthogonal. The rate of compression of ϱ' is the same as that for compressing ϱ in the $|\psi_j\rangle$ basis, which is at least the entropy of ϱ. Measurements make a system more random and increase the entropy of a the entropy of a system.

$$S_\varrho(A) \geq S(\varrho), \tag{5.31}$$

so that the entropy of the state as a whole is less than the entropy of any of its observables. Equality is achieved only when the observable commutes with the state (i.e. when the projections are in the diagonal basis of ϱ). This, in a way, can be interpreted as an instance of the increase in entropy in nature, which constitutes the basis for the Second Law of Thermodynamics. The entropy before the measurement is the von Neumann entropy of ϱ, and after the measurement the entropy of observation is higher—depending on the choice of the projective measurement. The difference between the two is the entropy increase. However, we shall see later on that this is not true for more general measurements, namely POVMs, where this difference can in fact be negative. Thus we shall have to be more careful in formulating the entropy changes during a measurement. This will be done in the next chapter, after I have introduced the full machinery of quantum information theory.

5.5 Conditional entropy and mutual information

The joint entropy of two classical systems is at least the entropy of one of the systems. A pure state of two systems has a von Neumann entropy of zero, yet when one of the systems is traced out, if the state is entangled, then the density operator obtained has a nonzero entropy. Thus the joint entropy of entangled quantum systems can be negative which shows is not possible in classical systems and shows how classical and quantum systems differ.[8] This is, in fact, is an entropic way of quantifying correlations, which we shall say more about later.

The relationships between the conditional, joint, and mutual entropies are shown in Fig. 5.2. The joint entropy of two quantum systems ϱ^A and ϱ^B is the von Neumann entropy of ϱ^{AB}, $S(\varrho^{AB})$. The conditional entropy is the information gained from ϱ^{AB} when ϱ^B is known:

$$S(\varrho^A|\varrho^B) = S(\varrho^{AB}) - S(\varrho^B). \tag{5.32}$$

$S(\varrho^B|\varrho^A)$ is defined analogously. The mutual information measures the correlations between ϱ^A and ϱ^B:

$$I(A:B) = S(A) + S(B) - S(A,B) \tag{5.33}$$
$$= S(A) - S(A|B) \tag{5.34}$$
$$= S(B) - S(B|A). \tag{5.35}$$

---

[8] Classically, the total entropy cannot be zero when its reductions are nonzero.

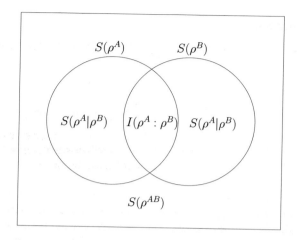

Fig. 5.2 The joint entropy of ϱ^{AB} is the entropy of the whole system. The mutual entropy measures the correlation between the systems. The conditional entropy between two systems is the entropy of the system minus any information gained from the other system from the correlations they share.

If ϱ^{AB} is pure, it can be expressed as its Schmidt decomposition,

$$\sum_i \alpha_i |\psi_i^A\rangle |\phi_i^B\rangle \ . \tag{5.36}$$

Its joint entropy $S(\varrho^{AB})$ is zero, but the entropy of one of the systems is

$$S(\varrho^A) = S(\varrho^B) = \sum_i |\alpha_i|^2 \log |\alpha_i|^2 \ , \tag{5.37}$$

which leads to the mutual entropy and the conditional entropy between A and B being

$$I(\varrho^A : \varrho^B) = 2S(\varrho^A) \tag{5.38}$$

and

$$S(\varrho^A|\varrho^B) = -S(\varrho^A) \ . \tag{5.39}$$

With classical states, this is not possible. The correlation between two random variables X and Y is bounded by the smaller of the reduced entropies $S(X)$ and $S(Y)$. The classical conditional entropy is also always positive:

$$I(X:Y) \leq H(X) , \tag{5.40}$$
$$H(X|Y) \geq 0 \ . \tag{5.41}$$

The fact that the quantum mutual information between two entangled systems in a pure state is equal to twice the reduced entropy while classically it can be at most one times the reduced entropy indicates to us that quantum correlations are stronger than classical.[9] Later chapters will be devoted entirely to showing precisely how to quantify this excess of quantum correlation (i.e. entanglement).

---

[9] As far as I am aware, this was first noted by Lindblad in 1973.

5.6 Relative entropy

The relative entropy between two density operators $\varrho = \sum_i p_i |\psi_i\rangle\langle\psi_i|$ and $\sigma = \sum_j q_j |\phi_j\rangle\langle\phi_j|$ is

$$S(\varrho||\sigma) = \text{tr}\,(\varrho \log \varrho - \varrho \log \sigma) \quad (5.42)$$

$$= \sum_i p_i \log p_i - \sum_{i,j} p_i \log q_j |\langle\psi_i|\phi_j\rangle|^2 \quad (5.43)$$

$$= -S(\varrho) - \sum_{i,j} p_i \log q_j |\langle\psi_i|\phi_j\rangle|^2 . \quad (5.44)$$

The relative entropy tells us how similar two density operators are. It is minimal when $\varrho = \sigma$, in which case $S(\varrho||\varrho) = 0$. The relative entropy increases as the distance between the probabilities p_i and q_i and the distance between the two bases $|\psi_i\rangle$ and $|\phi_i\rangle$ increase. It is maximal when the support of σ is orthogonal to that of ϱ, in which case $S(\varrho||\sigma) \to \infty$.

Remembering that the logarithm is a concave function (i.e. for any $\sum_i p_i = 1$ and any x, $\sum_i \log(p_i x) \le \log\left(\sum_i p_i x\right)$), we can show that

$$S(\varrho||\sigma) \le -\sum_i p_i \log p_i - \sum_i p_i \log \sum_j q_j |\langle\psi_i|\phi_j\rangle|^2 \quad (5.45)$$

$$= -\sum_i p_i \log p_i - \sum_{i,j} p_i \log q_i , \quad (5.46)$$

which is the classical relative entropy between two classical distributions with probabilities p_i and q_j. Equality holds only when both density operators are diagonalized in the same basis, that is, if they commute.

We shall now enumerate all of the important properties of the relative entropy that will be used in the rest of the book. It is quite staggering to see how many fundamental results in quantum information science follow from these simple relationships:

1. $S(\sigma||\sigma) = 0$. The distance from a state to itself is always zero.
2. $S(\sigma_1 \otimes \sigma_2 || \varrho_1 \otimes \varrho_2) = S(\sigma_1||\varrho_1) + S(\sigma_2||\varrho_2)$. The relative entropy inherits additivity from the von Neumann entropy.
3. $S(\lambda\sigma_1 + (1-\lambda)\sigma_2||\varrho) \le \lambda S(\sigma_1||\varrho) + (1-\lambda)S(\sigma_2||\varrho)$. This rule says that the relative entropy is convex: that is, mixing of physical states decreases the distance between them (and leads to indistinguishability).
4. $S(U\varrho U^\dagger || U\sigma U^\dagger) = S(\varrho||\sigma)$. The relative entropy, like the von Neumann entropy is invariant under unitary transformations of the states.
5. $S(\text{tr}_B \sigma || \text{tr}_B \varrho) \le S(\sigma||\varrho)$. A partial trace decreases distinguishability. The less information we have about two states (i.e. if we only have partial access to them), the less we can tell if there is any difference between them.
6. Donald's inequality: $\sum_k p_k S(\varrho_k||\sigma) = \sum_k p_k S(\varrho_k||\varrho) + S(\varrho||\sigma)$, where $\varrho = \sum_k p_k \varrho_k$. Donald's inequality simply states that the average distance to σ equals the average distance to ϱ plus the distance from ϱ to σ, where ϱ is the average state.

5.7 Statistical interpretation of relative entropy

The statistical value of the relative entropy is that it measures the closeness of two states as how easy they are to discriminate using the outcomes of measurements made on them. The higher the value of the relative entropy, the easier the states are to discriminate. This is true both if states are represented by classical probability distributions, or if we use density operators. The main result in this section, which we shall argue for, is that the probability of confusing a state ϱ with a state σ is (asymptotically, for N trials) given by

$$P_N(\varrho \to \sigma) = 2^{-NS(\sigma||\varrho)}. \tag{5.47}$$

A more operational interpretation of both the Shannon entropy and the relative entropy can be obtained from this statistical point of view. The generalization of this formalism to the quantum domain will be presented later. I now follow the approaches of Cover and Thomas (1991), and Csiszár and Körner (1981). The reader interested in more detail can consult these two books.

Let $X_1, X_2, \ldots, X_n$ be a sequence of n symbols drawn from an alphabet $A = \{a_1, a_2, \ldots, a_{|A|}\}$, where $|A|$ is the size of the alphabet. We denote a sequence $x_1, x_2, \ldots, x_n$ by x^n or, equivalently, by $\mathbf{x}$. The type $P_\mathbf{x}$ of a sequence $x_1, x_2, \ldots, x_n$ is defined as the relative proportion of occurrences of each symbol of A, that is, $P_\mathbf{x}(a) = N(a|\mathbf{x})/n$ for all $a \in A$, where $N(a|\mathbf{x})$ is the number of times the symbol a occurs in the sequence $\mathbf{x} \in A^n$. Thus, according to this definition, the sequences 011010 and 100110 are of the same type. $\mathcal{E}P_n$ will denote the set of types with denominator n, that is, the set of types of strings of length n. If $P \in \mathcal{E}P_n$, then the set of sequences of length n and type P is called the **type class** of P, denoted by $T(P)$, that is, mathematically,

$$T(P) = \{\mathbf{x} \in A^n : P_\mathbf{x} = P\}.$$

We now approach our first theorem about types, which is at the heart of the success of this theory, and states that the number of types increases only polynomially with n.

Theorem 1.

$$|\mathcal{E}P_n| \leq (n+1)^{|A|}.$$

The proof of this is left for the reader, but the rationale is simple. Suppose that we generate an n-bit string of 0's and 1's. The number of different types is then $n+1$, that is, polynomial in n. The zeroth type has only one string, all zeros; the first type has n strings, all strings containing exactly one 1; the second type has $n(n-1)/2$ strings, all those containing exactly two 1's; and so on. The nth type has only one sequence, all 1's. The most important point is that the number of sequences is exponential in n, so that at least one type has exponentially many sequences in its type class, since there are only polynomially many different types. A simple example is provided by a coin tossed n times. If it is a fair coin, then we expect heads half of the time and tails the other half of the time. The number of all possible sequences for this coin is 2^n (i.e. exponential in n) and each sequence is equally likely (with probability 2^{-n}). However, the size of the type class in which there are equal numbers of heads and tails is $C^n_{n/2}$ (the number of possible ways of choosing $n/2$ elements out of n elements), the logarithm of which tends to n for large n. Hence this type class is in some sense asymptotically as large as all the type classes together.

We now arrive at a very important theorem for us, which, in fact, presents the basis of the statistical interpretation of the Shannon entropy and the relative entropy.

Theorem 2. If $X_1, X_2, \ldots X_n$ are drawn according to the probability distribution $Q(x)$, then the probability of **x** depends only on its type and is given by

$$Q^n(\mathbf{x}) = 2^{-n(H(P_\mathbf{x}) + H(P_\mathbf{x} || Q))} .$$

Proof.

$$\begin{aligned} Q^n(\mathbf{x}) &= \prod_{i=1}^n Q(x_i) = \prod_{a \in A} Q(a)^{N(a|\mathbf{x})} \\ &= \prod_{a \in A} Q(a)^{n P_\mathbf{x}(a)} = \prod_{a \in A} 2^{n P_\mathbf{x}(a) \log Q(a)} \\ &= 2^{\{n \sum_{a \in A} -P_\mathbf{x}(a) \log(P_\mathbf{x}(a)/Q(a)) + P_\mathbf{x}(a) \log P_\mathbf{x}(a)\}} \\ &= 2^{-n(H(P_\mathbf{x}) + S(H_\mathbf{x} || Q))} .\square \end{aligned}$$

Therefore the probability of a sequence becomes exponentially small as n increases. Indeed, our coin tossing example shows this: the probability for any particular sequence (such as 0000011111) is 2^{-n}. This is explicitly stated in the following corollary.

Corollary. If **x** is in the type class of Q (i.e. the relative frequency of each character of **x** is given by the distribution Q), then the probability of the string **x** as n gets large is

$$Q^n(\mathbf{x}) = 2^{-nH(Q)} .$$

The proof of this is left to the reader.

So, when n becomes large, most of the sequences become typical and they are all equally likely. Therefore the probability of every typical sequence times the number of typical sequences has to be equal to 1 in order to conserve the total probability ($2^{-nH(Q)} N = 1$). From this we can see that the number of typical sequences is $N = 2^{nH(Q)}$ (we turn to this point more formally next). Hence, the above theorem has very important implications in the theory of statistical inference and of distinguishability of probability distributions. To see how this comes about, we shall now state two theorems that give bounds on the size and probability of a particular type class. The proofs follow directly from the above two theorems and the corollary.

Theorem 3. For any type $P \in \mathcal{E}P_n$,

$$\frac{1}{(n+1)^{|A|}} 2^{nH(P)} \leq |T(P)| \leq 2^{nH(P)}$$

This theorem provides exact bounds on the number of "typical" sequences (similarly to what we did in Chapter 1). Suppose that we have probabilities p_1 and p_2 for heads and tails respectively and we toss the coin n times. The typical (most likely) sequence will be the one where we have $p_1 n$ heads and $p_2 n$ tails. The number of such sequences is

$$C_{p_1 n}^n = \frac{n!}{(p_1 n)!(p_2 n)!} \sim 2^{n(-p_1 \log p_1 - p_2 \log p_2)},$$

that is, an exponential in n (more tosses, more possibilities) and in the entropy (higher uncertainty, more possibilities). The next theorem offers a statistical interpretation of the relative entropy.

Theorem 4. For any type $P \in \mathcal{E}P_n$ and any distribution Q, the probability of the type class $T(P)$ when the probability of a string is determined by Q^n is $2^{-nH(P||Q)}$ to the first order in the exponent. More precisely,

$$\frac{1}{(n+1)^{|A|}} 2^{-nH(P||Q)} \leq Q^n(T(P)) \leq 2^{-nH(P||Q)}.$$

The meaning of this theorem is that if we draw results according to the distribution Q, the probability that they will "look" as if they were drawn from P decreases exponentially with n and with the relative entropy between P and Q. The closer Q is to P, the higher the probability that their statistics will look the same. Alternatively, the higher the number of draws n, the smaller the probability that we shall confuse the two. We present an explicit example below. The above two results can be written succinctly in an exponential fashion that will be useful to us, as follows:

$$|T(P)| \rightarrow 2^{-nH(P)}, \qquad (5.48)$$
$$Q^n(T(P)) \rightarrow 2^{-nH(P||Q)}. \qquad (5.49)$$

The first statement also leads to the idea of data compression, where the typical strings of length n generated by a source with entropy H can be encoded into strings of length nH. The second statement says that if we are performing n experiments according to a distribution Q, the probability that we shall get something that looks as if it was generated by a distribution P decreases exponentially with n, depending on the relative entropy between P and Q. This idea leads immediately to Sanov's theorem, whose quantum analogue will provide a statistical interpretation of a measure of entanglement. Now we present some examples of data compression and introduce Sanov's theorem.

Let us we look at the distinguishability of two probability distributions. Suppose we would like to check if a given coin is "fair", that is, if it generates a "head–tail" distribution of $f = (1/2, 1/2)$. If the coin is biased, it will produce some other distribution, say $uf = (1/3, 2/3)$. So, our question of the fairness of the coin boils down to how well we can differentiate between two given probability distributions given a finite number n of experiments to be performed on one of the two distributions. In the case of the coin, we would toss it n times and record the number of 0's and 1's. From simple statistics we know that if the coin is fair then the number of 0's, $N(0)$, will be roughly in the range $n/2 - \sqrt{n} \leq N(0) \leq n/2 + \sqrt{n}$ for large n, and the same will be true for the number of 1's. So if our experimentally determined values do not fall within the above limits, the coin is not fair. We can look at this from another point of view, which is in the spirit of the method of types: namely, what is the probability that a fair coin will be mistaken for an unfair one with a distribution of $(1/3, 2/3)$ given n trials of the fair coin? For large n, we already know that

$$p(\text{fair} \to \text{unfair}) = 2^{-nH(uf||f)},$$

where $H(uf||f) = (1/3)\log(1/3) + (2/3)\log(2/3) - (1/3)\log(1/2) - (2/3)\log(1/2)$ is the classical relative entropy for the two distributions. So,

$$p(\text{fair} \to \text{unfair}) = 3^n 2^{-(5/3)n},$$

which tends exponentially to zero as $n \to \infty$. In fact, we can see that after ~ 20 trials, the probability of mistaking one distribution for another is vanishingly small, $< 10^{-10}$. This leads to the following important result:

Sanov's theorem. If we have a probability distribution Q and a set of distributions $E \subset \mathcal{E}P$, then

$$Q^n(E) \to 2^{-nH(P^*||Q)}, \tag{5.50}$$

where P^* is the distribution in E that is closest to Q using the Shannon relative entropy.

This can also be rephrased in the language of distinguishability: when we are trying to distinguish a given distribution from a set of distributions, then what matters is how well we can distinguish that distribution from the closest one in the set. When we turn to the quantum case, as we do next, the probability distributions will become quantum densities representing various states of a quantum system, and the question will be how well we can distinguish between these states. Note that we could also talk about Q coming from a set of states in which case we would have $H(P||Q^*)$, Q^* being the state that minimizes the relative entropy (i.e. the closest state).

The quantum relative entropy has the same statistical interpretation as its classical analogue: it tells us how difficult it is to distinguish a state σ from the state. To this end, suppose we have states σ and ϱ. How can we distinguish them? We can chose a POVM $\sum_{i=1} M_i = I$, which generates two distributions via

$$p_i = \operatorname{tr} M_i \sigma, \tag{5.51}$$
$$q_i = \operatorname{tr} M_i \varrho, \tag{5.52}$$

and use classical reasoning to distinguish these two distributions. However, the choice of POVMs is not unique. It is therefore best to choose that POVM which distinguishes the distributions most, that is, for which the *classical* relative entropy is largest. Thus we arrive at the following quantity:

$$S_1(\sigma||\varrho) := \sup_{M's} \left\{ \sum_i \operatorname{tr} M_i \sigma \log \operatorname{tr} M_i \sigma - \operatorname{tr} M_i \sigma \log \operatorname{tr} M_i \varrho \right\}, \tag{5.53}$$

where the supremum is taken over all POVMs. The above is not the most general measurement that we can make, however. In general we have N copies of σ and ϱ in the state:

$$\sigma^N = \underbrace{\sigma \otimes \sigma \ldots \otimes \sigma}_{\text{total of } N \text{ terms}} \quad (5.54)$$

$$\varrho^N = \underbrace{\varrho \otimes \varrho \ldots \otimes \varrho}_{\text{total of } N \text{ terms}} \quad (5.55)$$

We may now apply a POVM acting on σ^N and ϱ^N. Consequently, we define a new type of relative entropy,

$$S_N(\sigma||\varrho) := \sup_{\text{M's}} \{\frac{1}{N} \sum_i \text{tr} M_i \sigma^N \log \text{tr} M_i \sigma^N - \text{tr} M_i \sigma^N \log \text{tr} M_i \varrho^N\} \quad (5.56)$$

Now it can be shown that

$$S(\sigma||\varrho) \geq S_N(\sigma||\varrho) , \quad (5.57)$$

where $S(\sigma||\varrho)$ is the quantum relative entropy. (This really is a consequence of the fact that the relative entropy does not increase under general CP-maps, a fact that will be proven formally in the next chapter). Equality is achieved in eqn 5.57 if and only if σ and ϱ commute. However, for any σ and ϱ, it is true that

$$S(\sigma||\varrho) = \lim_{N \to \infty} S_N(\sigma||\varrho) . \quad (5.58)$$

In fact, this limit can be achieved by projective measurements which are independent of σ. From these considerations it naturally follows that the probability of confusing two quantum states σ and ϱ (after performing N measurements on ϱ) is (for large N)

$$P_N(\varrho \to \sigma) = e^{-NS(\sigma||\varrho)} . \quad (5.59)$$

We would like to stress here that classical statistical reasoning applied to the problem of distinguishing quantum states leads to the above formula. There are, however, other approaches. Some take eqn 5.59 for their starting point and then derive the rest of the formalism from that. Others, on the other hand, assume a set of axioms that must be satisfied by the quantum analogue of the relative entropy (e.g. it should reduce to the classical relative entropy if the density operators commute, i.e. if they are "classical") and then derive eqn 5.59 as a consequence. In any case, as we have argued here, there is a strong reason to believe that the quantum relative entropy $S(\sigma||\varrho)$ plays the same role in quantum statistics as the classical relative entropy plays in classical statistics.

5.8 Summary

The von Neumann entropy is a unique function on quantum states (up to an affine transformation) which is purely a function of probability, continuous, and additive. The von Neumann entropy of a density operator $\varrho = \sum_i p_i |i\rangle\langle i|$, where the $|i\rangle$'s are an eigenbasis (i.e. orthogonal), is defined by any of the following equivalent definitions:

$$S(\varrho) = -\text{tr}(\varrho \log(\varrho)) , \quad (5.60)$$

$$S(\varrho) = -\sum_i p_i \log(p_i) , \quad (5.61)$$

$$S(\varrho) = H(p_i) . \quad (5.62)$$

The fidelity F between two pure states is defined as

$$F(|\psi\rangle, |\phi\rangle) = |\langle\phi|\psi\rangle|^2 , \qquad (5.63)$$

which is the probability of measuring $|\psi\rangle$ as $|\phi\rangle$ or vice versa. Depending on the author, the fidelity F between two mixed states may be defined as the probability of measuring one as the other

$$F(\varrho, \sigma) = \mathrm{tr}(\varrho\sigma) . \qquad (5.64)$$

Some authors, however, prefer to have $F(\varrho, \varrho) = 1$, in which case they define the fidelity as

$$F(\varrho, \sigma) = \mathrm{tr}(\sqrt{\varrho\sigma}) . \qquad (5.65)$$

For the purposes of this book, it is unimportant which fidelity measure is used.

Helstrom's discrimination problem is to distinguish between two pure states $|\psi\rangle$ and $|\phi\rangle$. The optimal strategy is to choose a projective measurement which is equally likely to correctly distinguish $|\psi\rangle$ from $|\phi\rangle$ and vice versa. The optimal probability of success is

$$p_S = \frac{1}{2}(1 + \sqrt{1 - |\langle\psi_1|\psi_2\rangle|}) . \qquad (5.66)$$

The von Neumann entropy of a mixture of quantum states is the optimal rate of compression. If we write the density operator of the mixture in diagonal form,

$$\varrho = \sum_i p_i |i\rangle\langle i| , \qquad (5.67)$$

the analysis of quantum data compression largely follows the analysis of the Shannon entropy for compressing the probability distribution $P(X = i) = p_i$. However, with quantum data compression, we need to show also that the fidelity between the compressed state ϱ' and the original state ϱ is arbitrarily high.

The entropy of observation is the entropy of a set of observations made on a quantum system. Projective measurements always increase entropy. We can define mutual information, conditional entropy, and joint entropy in the same way as they are defined classically (using the diagonal basis). When a quantum state is entangled, the mutual information and the conditional entropies between its parts exceed the bounds of their classical equivalents owing to nonclassical correlations.

Statistically, the relative entropy expresses the asymptotic probability of confusing two probability distributions (in classical information theory) or two density matrices (in quantum information theory). The fact that the entropy gives a limit to the extent to which data can be compressed follows from this result both in classical and in quantum information theory.

6
Quantum information II

We now start the more formal exposition of the principles of quantum information. We shall, in this chapter, obtain the quantum mechanical equivalent of the Shannon noisy-channel communication theorem, as well as some profound statements about the behavior of quantum information during generalized quantum measurements. The latter will be very important in our studies of quantum entanglement, but are also significant in exploring the connections between thermodynamics, information theory and quantum physics. We shall have much more to say about this later in the book.

First we present some useful inequalities that will be instrumental in proving quantum analogues of Shannon's classical results (Ingarden, Kossakowski and Ohya, 1997). Where appropriate, proofs will be presented; otherwise, intuitive arguments will suffice.

6.1 Equalities and inequalities related to entropy

The relative entropy and mutual information are linked by the formula

$$S(\varrho^{AB}||\varrho^A \otimes \varrho^B) = S(\varrho^{AB}) - S(\varrho^A) - S(\varrho^B) = I(\varrho^A : \varrho^B) \,, \tag{6.1}$$

which says that the relative entropy between a joint system and its two parts is the same as the correlations between the two parts.[1] Here is a simple proof:

$$\begin{align}
S(\varrho^{AB}||\varrho^A\varrho^B) &= S(\varrho^{AB}) + \mathrm{tr}(\varrho^{AB}\log(\varrho^A \otimes \varrho^B)) \tag{6.2}\\
&= S(\varrho^{AB}) + \mathrm{tr}(\varrho^{AB}\log(\varrho^A)) + \mathrm{tr}(\varrho^{AB}\log(\varrho^B)) \tag{6.3}\\
&= S(\varrho^{AB}) - S(\varrho^A) - S(\varrho^B) \,. \tag{6.4}
\end{align}$$

An immediate consequence of this result is the subadditivity theorem. The relative entropy is always positive so $S(\varrho^{AB}||\varrho^A \otimes \varrho^B) \geq 0$ and we have

$$S(\varrho^{AB}) \geq S(\varrho^A) + S(\varrho^B) \tag{6.5}$$

The subadditivity theorem says that no matter how correlated two systems are, the information contained in the whole is at most the sum of the information contained in the parts. The message here is that, owing to correlations, the mixedness of the total state is in general smaller than the mixedness of the individual parts.

---

[1] In other words, the mutual information between the joint system AB and system A (with B traced out) and system B (with A traced out) is equal to the mutual information between the systems A and B.

The subadditivity theorem can be used to prove the triangular inequality. Let ϱ^{AB} be purified to ϱ^{ABC} (meaning that $\mathrm{tr}_C \varrho^{ABC} = \varrho^{AB}$). Using the Schmidt decomposition (we can use this because ϱ^{ABC} is pure) on the systems $\varrho^{AC} \otimes \varrho^B$ and $\varrho^{AC} \otimes \varrho^B$, we have

$$S(\varrho^C) = S(\varrho^{AB}) \qquad (6.6)$$
$$, S(\varrho^{AC}) = S(\varrho^B). \qquad (6.7)$$

Using subadditivity, we get

$$\begin{aligned} S(\varrho^B) &= S(\varrho^{AC}) & (6.8) \\ &\leq S(\varrho^A) + S(\varrho^C) & (6.9) \\ &= S(\varrho^A) + S(\varrho^{AB}). & (6.10) \end{aligned}$$

By exchanging the systems A and B, we get the triangular inequality which bounds the entropy of the two systems,

$$S(\varrho^{AB}) \geq |S(\varrho^A) - S(\varrho^B)|. \qquad (6.11)$$

This is an important bound, and for pure states it immediately shows that the reduced entropies of A and B are the same (as we concluded from the Schmidt decomposition earlier).

I now present one of the most important theorems in the book, which will have a huge bearing on many subsequent results. We shall be able to derive the capacity of a quantum communication channel from it, as well as various measures of entanglement.

Theorem. The relative entropy never increases under CP-maps:

$$S(\sigma||\varrho) \geq S(\Phi(\sigma)||\Phi(\varrho)),$$

where Φ is the most general CP-map.[2]

I shall first present a physical argument as to why we should expect this theorem to hold. As I have discussed, a CP-map can be represented as a unitary transformation on an extended Hilbert space. Unitary transformations do not change the relative entropy between two states simply because the entropy itself is invariant under unitary transformations. However, after this, we have to perform a partial trace to go back to the original Hilbert space, which, as is intuitively clear, decreases the relative entropy as some information is invariably lost during this operation. Hence the relative entropy decreases under any CP-map (see Fig. 6.1). I now formalize this proof.

Proof. I have discussed the fact that a CP-map can always be represented as a unitary operation together with a partial trace on an extended Hilbert space $\mathcal{E}H \otimes \mathcal{E}H_n$, where $\dim \mathcal{E}H_n = n$ (the treatment I shall follow is by Lindblad, who, I believe, gave the

---

[2] A nice colloquial version of this theorem is "things can only get worse" meaning that things become less distinguishable as a consequence of a general physical evolution. It is amazing to see how many other results follow from this rather pessimistic statement.

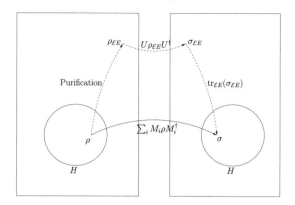

Fig. 6.1 Two ways of considering a CP-map. The first is that ϱ is transformed to σ by a CP-map, i.e. $\sigma = \sum_i M_i \varrho M_i^\dagger$. The second is that ϱ is purified to $\varrho_{\mathcal{E}E}$ in the extended Hilbert space $\mathcal{E}H \otimes \mathcal{E}H_n$. $\varrho_{\mathcal{E}E}$ can be thought of as ϱ and any parts of the environment with which it is entangled, and ϱ represents the part of $\varrho_{\mathcal{E}E}$ in the system at hand. A unitary operation U acts on $\varrho_{\mathcal{E}E}$, which leaves the state $\sigma_{\mathcal{E}E}$. The environment can now be traced out, leaving σ.

first clear account of this). Let $\{|i\rangle\}$ be an orthonormal basis in $\mathcal{E}H_n$ and let $|\psi\rangle$ be a unit vector. We define

$$W = \sum_i V_i \otimes |i\rangle\langle\psi| , \qquad (6.12)$$

Now, $W^\dagger W = I \otimes P_\psi$, where $P_\psi = |\psi\rangle\langle\psi|$, and there is a unitary operator U in $\mathcal{E}H \otimes \mathcal{E}H_n$ such that $W = U(I \otimes P_\psi)$ (this is a very well-known property, which can, for example, be found in Reed and Simon, 1980). Consequently,

$$U(A \otimes P_\psi)U^\dagger = \sum_{ij} V_i A V_j^\dagger \otimes |i\rangle\langle j| \qquad (6.13)$$

Tracing out the system $\mathcal{E}H_n$, we obtain

$$\mathrm{tr}_2\{U(A \otimes P_\psi)U^\dagger\} = \sum_i V_i A V_i^\dagger$$

This shows that the unitary representation and the representation by $\sum_i V_i \varrho V_i^\dagger$ are equivalent.

Now we can set up a chain of inequalities as follows:

$$\begin{aligned}
S(\Phi(\sigma)||\Phi(\varrho)) &= S(\mathrm{tr}_2\{U(\sigma \otimes P_\psi)U^\dagger\}||\mathrm{tr}_2\{U(\varrho \otimes P_\psi)U^\dagger\}) \\
&\leq S(U(\sigma \otimes P_\psi)U^\dagger||U(\varrho \otimes P_\psi)U^\dagger) \\
&= S(\sigma \otimes P_\psi||\varrho \otimes P_\psi) \\
&= S(\sigma||\varrho) + S(P_\psi||P_\psi) \\
&= S(\sigma||\varrho) .
\end{aligned}$$

The first line follows from the fact that any CP-map can be written as a unitary transformation on an extended Hilbert space. The second line follows from the fact that partial tracing cannot increase the relative entropy. This, in turn, is a consequence of the concavity of the log function. The third line follows from the fact that the relative entropy is invariant under unitary transformations. The fourth line is just a consequence of the additivity of relative entropy, and the last line follows from the fact that the relative entropy between a state and itself vanishes. This proves the result that the relative entropy cannot increase under a CP-map □.

A simple consequence of the fact that the quantum relative entropy does not increase under CP-maps is that correlations (as measured by the quantum mutual information) also cannot increase, but this now applies under *local* CP-maps. This is a very satisfactory property of a measure of correlation: it says, somehow, that two systems cannot increase their correlation without in some sense interacting, or exchanging information.

6.2 The Holevo bound

Let Alice now encode her messages into general mixed states ϱ_i with respective probabilities p_i, and then send these to Bob for discrimination. We can think of this as a model of a noisy quantum channel: we can say that Alice has encoded her messages into pure states ψ_i, but that these have become mixed when sent through the noisy channel. Bob now has to discriminate between mixed states rather than the original pure states and we therefore expect his capacity to be reduced (i.e. it will be less than the von Neumann entropy of the mixture of messages $\varrho = \sum_i p_i \varrho_i$). But the question is by how much?

Let ϱ be a quantum system prepared in state ϱ_i with probability p_i, and let the system then be measured by use of the measurement operators $\{M_j\}_j$ with $\sum_j M_j = I$. Let X be a random variable, where $P(X = i) = p_i$ and let Y be a random variable with $P(Y = j)$, which is the probability that j is obtained as the outcome of a measurement on system ϱ. The Holevo bound states that the mutual information between the preparation probability $P(X = x)$ and the measurement probability is bounded by

$$I(X:Y) \leq S(\varrho) - \sum_i p_i S(\varrho_i) \,. \tag{6.14}$$

The right-hand side is often called the Holevo χ quantity,

$$\chi(\varrho) = S(\varrho) - \sum_i p_i S(\varrho_i) \,, \tag{6.15}$$

and represents the difference between the entropy of the whole and the average entropy of the parts. It plays the same role in the theory of quantum communication as classical mutual information does in Shannon's information theory.

Proof of the Holevo bound. The Holevo bound is a direct consequence of the fact that the quantum relative entropy does not increase under CP-maps.[3] Here we use a simpler proof than the original, which basically entails applying the fact that relative entropy does not increase under CP-maps (twice). One such CP-map is given by

$$\tau(A) = \frac{1}{n}\text{tr}(A) \,,$$

[3] Holevo's original proof is much more complicated and does not use the quantum relative entropy.

where A is any $n \times n$ positive matrix. When applied to the relative entropy, this leads to the well-known Pierls–Bogoliubov inequality (PBI), (Bhatia 1997)

$$\tau(A)(\log \tau(A) - \log \tau(B)) \leq \tau(A \log A - A \log B) . \tag{6.16}$$

To prove the Holevo bound, I again use that fact that the relative entropy does not increase under CP-maps, to write

$$S(\varrho_i || \varrho) \geq \sum_j S(A_j \varrho_i A_j^\dagger || A_j \varrho A_j^\dagger) .$$

The PBI now implies that

$$\begin{aligned} S(A_j \varrho_i A_j^\dagger || A_j \varrho A_j^\dagger) &\geq \operatorname{tr}(A_j \varrho_i A_j^\dagger) \{ \log(\operatorname{tr}(A_j \varrho_i A_j^\dagger)) - \log(\operatorname{tr}(A_j \varrho A_j^\dagger)) \} \\ &= p(j|i)(\log p(j|i) - \log p(j)) ,\end{aligned}$$

where $p(j|i) = \operatorname{tr}\{A_j \varrho_i A_j^\dagger\}$ is the conditional probability that the message ϱ_i will lead to the outcome $E_j = A_j^\dagger A_j$, and $p(j) = \sum_i p(j|i)$. Thus we now have that

$$S(\varrho_i || \varrho) \geq \sum_j p(j|i)(\log p(j|i) - \log p(j)) .$$

Multiplying both sides by p_i (which is positive) and summing over all i leads to the Holevo bound. This is the most elegant proof of the Holevo bound that I am aware of.[4] □

Since Holevo's result is one of the key results in quantum information theory, I shall present another simple way of understanding it using the quantum mutual information. This, of course, is only an additional motivation for believing the Holevo bound, and by no means proves its validity. Namely, if Alice encodes the symbol (Sym) i into the state (St) ϱ_i, then the total state (Sym + St) is

$$\varrho_{\text{Sym+St}} = \sum_i p_i |i\rangle\langle i| \otimes \varrho_i ,$$

where the kets $|i\rangle$ are orthogonal (we can think of these as representing different states of consciousness of Alice!). Bob now wants to learn about the symbols by distinguishing the states ϱ_i. He cannot learn more about the symbols than is already stored in the correlations between the symbols and the message states. These correlations, as we know, are measured by the quantum mutual information

$$\begin{aligned} I(\varrho_{\text{Sym+St}}) &= S(\text{Sym}) + S(\text{St}) - S(\text{Sym + St}) \\ &= S\left(\sum_i p_i \varrho_i\right) - \sum_i p_i S(\varrho_i) , \end{aligned} \tag{6.17}$$

which is the same as the Holevo bound.

I would like now to derive the capacity of a classical communication channel from the Holevo bound.[5] As I mentioned before, the Holevo bound itself contains the classi-

---

[4] The Holevo bound is, in some sense, obvious! It says that the quantum distinguishability between the initial states used by Alice (given by the average relative entropy) cannot be increased by making measurements on them (performed by Bob in our scenario).

[5] I follow Gordon's reasoning who was, in fact, the first person to conjecture the Holevo bound (back in 1964!).

cal capacity of a classical channel as a special case. This, as we might expect, happens when all ϱ_i's are diagonal in the same basis, that is, they commute (classically, all states and observables commute because they can be simultaneously specified and measured, which is in contrast to quantum mechanics). Therefore density matrices are reduced to classical probability distributions. Let us call this basis the B representation, with orthonormal eigenvectors $|b\rangle$. The probability that a measurement of the symbol represented by ϱ_i will yield the value b is just $\langle b|\varrho_i|b\rangle$. This I call the conditional probability $p_i(b) = p(i|b)$, the probability that if ϱ_i was sent the result b was obtained. Now the Holevo bound is

$$C = S(\varrho) - \sum_i p_i S(\varrho_i) = S(\varrho) - S_B(\varrho_i),$$

where $S_B(\varrho_i)$ is the conditional entropy, given by

$$S_B(\varrho_i) = \sum_i p_i \sum_b \langle b|\varrho_i|b\rangle \log \langle b|\varrho_i|b\rangle = \sum_i p_i S(\varrho_i).$$

Thus, the Holevo bound reduces to the Shannon mutual information between the commuting messages and the measurement in the B representation.

In general, the usual rule of thumb for obtaining quantum-information-theoretic quantities from their classical counterparts is by the convention[6]

$$\sum \longrightarrow \text{Trace},$$
$$\sum p(a) \longrightarrow \varrho_A,$$

so that, for example, the Shannon entropy $S(p(a)) = -\sum_i p(a_i) \log p(a_i)$ now becomes the von Neumann entropy $S(\varrho_A) = -\text{tr}\, \varrho_A \log \varrho_A$.

6.3 Capacity of a bosonic channel

As the first application of the Holevo bound, I shall compute the channel capacity of a bosonic field, for example an electromagnetic field (for an excellent review, see Caves and Drummond 1994). The message information is assumed to be encoded into modes of frequency ω and average photon number $\bar{m}(\omega)$. The signal power is assumed be S. The noise in the channel is quantified by the average number of excitations $\bar{n}(\omega)$ and is assumed to be independent of the signal (i.e. the powers of the signal and of the noise are additive). We have seen that when there is no noise in the channel, the Holevo bound is equal to the entropy of the average signal. In order to compute the capacity, we need to maximize this entropy with the constraint that the total power (or energy) is fixed. It is well-known that thermal states are states that maximize the entropy. We thus assume that both the noise and the signal + noise are in thermal equilibrium and follow the usual Bose–Einstein statistics. The noise power is given by a standard formula from thermodynamics,

$$N = \frac{\pi (kT)^2}{12\hbar}.$$

---

[6]I must admit that I have been able to derive quite a lot of results in the field of quantum information from the theory of classical information using this quick and dirty substitution.

The power of the output from the channel (signal + noise) is

$$P = S + N = \frac{\pi(kT_e)^2}{12\hbar},$$

where T_e is the equilibrium temperature of the signal + noise. Therefore it follows that

$$T_e = \left(\frac{12\hbar S}{\pi k^2 + T^2}\right)^{1/2}.$$

The state of the noise in the mode ω is given by the Bose distribution,

$$\varrho_N(\omega) = \sum_n \frac{1 - e^{-\hbar\omega/kT}}{e^{\bar{n}(\omega)\hbar\omega/kT}} |n\rangle\langle n|,$$

and the state of the output is

$$\varrho_{N+S}(\omega) = \sum_n \frac{1 - e^{-\hbar\omega/kT_e}}{e^{\bar{n}(\omega)\hbar\omega/kT_e}} |n\rangle\langle n|$$

The capacity of the channel is given by the Holevo bound, which is

$$\begin{aligned} C &= \int_{-\infty}^{\infty} [S(\varrho_{S+N}(\omega)) - S(\varrho_N(\omega))]\, d\omega \quad (6.18) \\ &= \frac{\pi kT}{6\hbar \log 2} \left\{ \frac{12\hbar S}{\pi(kT)^2 + 1}^{1/2} - 1 \right\}. \end{aligned}$$

The integration is there to take into account all the modes of the field.

Let us look at the two extreme limits of this capacity. In the high-temperature limit we obtain the "classical" capacity,[7]

$$C_C = \frac{S}{kT \log 2}. \quad (6.19)$$

This states that in order to communicate one bit of information with this setup, we need an amount of energy of exactly $kT \ln 2$. In the low-temperature limit, on the other hand, quantum effects become important and the capacity becomes independent of T:[8]

$$C_Q = \frac{\sqrt{\pi}}{\log 2} \left(\frac{S}{\hbar}\right)^{1/2}. \quad (6.20)$$

Note the appearance of Planck's constant, which is a key feature of quantum mechanics. If we wish to communicate one bit of information in this limit, we need only

---

[7] A result derived by Shannon and Weaver (1949).

[8] This result was derived by Stern (1960), Gordon (1964), Lebedev and Levitin (1963) and Yamamoto and Haus (1986).

$\hbar/\pi(\ln 2) \sim 10^{-34}$J of energy. This is significantly less than the corresponding energy in the classical limit. Let us now compare the classical and quantum capacity limits with the total energy of N harmonic oscillators (bosons) in the same two limits. In the high-temperature limit the equipartition theorem is applicable and the total energy is $3NkT$ (i.e. it depends on temperature). In the low-temperature limit all the harmonic oscillators settle down to the ground state, so that the total energy becomes $N\hbar\omega/2$ (i.e. it is independent of temperature and we see the quantum dependence through Planck's constant $\hbar$).

6.4 Information gained through measurements

Suppose Alice prepares a mixture of nonorthogonal pure states $\varrho = \sum_i p_i |\psi_i\rangle\langle\psi_i|$ according to some classical probability distribution X with $P(X = i) = p_i$. She hopes that she can encode extra classical information by using nonorthogonal states. The entropy of X, $H(X)$, is called the entropy of preparation. Without telling Bob what the state is, she sends ϱ to Bob. He then measures ϱ by applying any measurement operators M_j he likes (subject to the completeness condition) and obtains an outcome j with probability $P(Y = j)$. The maximum amount of information that he can learn about the original probability distribution X is the mutual information $I(X : Y)$. The Holevo bound says that

$$I(X:Y) \leq S(\varrho) - \sum_i p_i S(|\psi_i\rangle\langle\psi_i|) \tag{6.21}$$

$$= S(\varrho). \tag{6.22}$$

This implies that the maximum amount of information that we can a encode in a mixture is its von Neumann entropy $S(\varrho)$.

We said in the last chapter that the von Neumann entropy of a mixture of pure states is the maximal compression rate of that mixture. We can now see that the von Neumann entropy also represents the maximal amount of classical information that can be extracted from a mixture of pure states.

Now Alice can try to encode extra classical information by encoding each i as a mixture ϱ_i. She prepares a mixture of mixtures $\varrho = \sum_i p_i \varrho_i$ to give to Bob as above; again, the maximal amount of information Bob can learn is bounded by

$$I(X:Y) \leq S(\varrho) - \sum_i p_i S(\varrho_i). \tag{6.23}$$

Plugging in

$$S(\varrho) \leq \sum_i p_i S(\varrho_i) + H(p_i), \tag{6.24}$$

we obtain

$$I(X:Y) \leq H(X). \tag{6.25}$$

This tells us that if Alice wants to send Bob information encoded in a random variable X, she can do no better than to encode the information into orthogonal states. And this corresponds to the classical communication capacity.

6.5 Relative entropy and thermodynamics

The Second Law of Thermodynamics states that the entropy of an isolated system never decreases.[9] This does not follow directly from the theorem that quantum relative entropy does not increase under CP-maps. Strictly speaking, an isolated system in quantum mechanics evolves unitarily and therefore its entropy never changes. Under a CP-map, on the other hand, the entropy can both increase and decrease. If, however, the state ϱ is maximally mixed, I/n, for example, then the quantum relative entropy is given by

$$S(\sigma||\varrho) = \log n - S(\sigma) . \qquad (6.26)$$

If in addition the evolution is such that I/n is a stationary state (i.e. invariant under the CP-map used), then the monotonic decrease in the quantum relative entropy implies a monotonic increase in $S(\sigma)$, just as in the Second Law of Thermodynamics. A detailed discussion of the statistical foundations of the Second Law can be found in the classic text Tollman (1938).

We have seen that communication essentially creates correlations between the sender and the receiver. Creating correlations is therefore very important if one is to be able to convey any information. However, I would now like to talk about the opposite process—destroying correlations. Why would one want to do this? The reason is that we might want to correlate one system with another and might need to delete all its previous correlations to be able to store new ones. I would like to give a more physical statement about information erasure and link it to the notion of measurement. I shall therefore introduce two correlated parties—a system and an apparatus. The apparatus will interact with the system, thereby gaining a certain amount of information about it. Suppose now that the apparatus needs to measure another system. We first need to delete information about the last system before we can make another measurement. The most general way of conducting erasure (resetting) of the apparatus is by employing a reservoir in thermal equilibrium at a certain temperature T. To erase the state of the apparatus we just throw it into the reservoir and introduce a new pure state. The entropy increase resulting from the operation now consists of two parts. First, the state of the apparatus evolves to the state of the reservoir, and the resulting entropy is now added to the entropy of the reservoir. Second, the rest of the reservoir changes its entropy because of this interaction, this change in entropy is equal to the difference in the apparatus internal energy before and after the resetting (no work is done in this process). A good model is obtained by imagining that the reservoir consists of a large number of systems (of the same "size" as the apparatus) all in the same quantum equilibrium state ω. The apparatus, which is in some state ϱ, interacts with these reservoir systems one at a time. Each time there is an interaction, the state of the apparatus approaches more closely the state of the reservoir, while that the individual system in the reservoir also changes its state away from the equilibrium. However, the systems in the bath are numerous so that after

---

[9] The whole point of linking our considerations so far to the Second Law is that the Second Law is a principle of great generality; it can be used to rule out a great number of processes otherwise consistent with other laws of physics. We shall see that the Second Law can be used for a very quick derivation that von Neumann entropy is the optimal rate of quantum data compression, as well as some more general statements about quantum measurement (Brillouin 1956).

a certain number of "collisions" the state of the apparatus will approach the state of the reservoir, whereas the state of the reservoir will not change much since it is very large.

Bearing all this in mind, we now reset the apparatus by plunging it into a reservoir in thermal equilibrium (in a Gibbs state) at a temperature T. Let the state of the reservoir be

$$\omega = \frac{e^{-\beta H}}{Z} = \sum_j q_j \ket{\varepsilon_j}\bra{\varepsilon_j}, \qquad (6.27)$$

where $H = \sum_i \varepsilon_i \ket{\varepsilon_i}\bra{\varepsilon_i}$ is the Hamiltonian of the reservoir, $Z = \mathrm{tr}(e^{-\beta H})$ is the partition function, and $\beta^{-1} = kT$, where k is the Boltzmann constant. Now suppose that owing to the measurement, the entropy of the apparatus is $S(\varrho)$ (and an amount $S(\varrho)$ of information has been gained), where $\varrho = \sum_i r_i \ket{r_i}\bra{r_i}$ is the eigenexpansion of the state of the apparatus. Now the total entropy increase during the erasure is the following (there are two parts, as I have argued above: (1) the change in the entropy of the apparatus, and (2) the change in the entropy of the reservoir):

$$\Delta S_{\mathrm{er}} = \Delta S_{\mathrm{app}} + \Delta S_{\mathrm{res}}. \qquad (6.28)$$

We know immediately that $\Delta S_{\mathrm{app}} = S(\omega)$, since the state of the apparatus (no matter what state it was in before) has now been erased so that it is the same as that of the reservoir. On the other hand, the entropy change in the reservoir is the average over all states $\ket{r_i}$ of the quantity of heat received by the reservoir divided by the temperature. This is the negative of the quantity of heat received by the apparatus divided by the temperature; the quantity of heat received by the apparatus is the internal energy after the resetting, minus the initial internal energy $\bra{r_i} H \ket{r_i}$. Thus,

$$\begin{aligned}
\Delta S_{\mathrm{res}} &= -\sum_k r_k \frac{\mathrm{tr}(\omega H) - \bra{r_k} H \ket{r_k}}{T} \\
&= \sum_k \left(r_k \sum_j |\braket{r_k|\varepsilon_j}|^2 - q_k \right)(-\log q_k - \log Z) \\
&= -\mathrm{tr}(\varrho - \omega)(\log \omega - \log Z) = \mathrm{tr}(\omega - \varrho)\log \omega.
\end{aligned}$$

Altogether, we now have an exact expression for the entropy increase due to deletion as below.

6.6 Entropy increase due to erasure

The entropy increase due to erasure is

$$\Delta S_{\mathrm{er}} = -\mathrm{tr}(\varrho \log \omega). \qquad (6.29)$$

In general, the information gain is equal to $S(\varrho)$, the entropy increase in the apparatus. This entropy increase is a maximum; the mutual information between the system and the apparatus is usually smaller, as in eqn 6.17. Thus, we see that

$$\Delta S_{\mathrm{er}} = -\mathrm{tr}(\varrho \log \omega) \geq S(\varrho) = I \qquad (6.30)$$

and this then confirms what is known as Landauer's erasure principle (the inequality follows from the fact that the quantum relative entropy $S(\varrho || \omega) = -\mathrm{tr}(\varrho \log \omega) - S(\varrho)$

is nonnegative). This principle states that the entropy increase due to the erased information has to be at least as big as the amount of information itself.

The erasure is least wasteful when $\omega = \varrho$, in which case the entropy of erasure is equal to $S(\varrho)$, the information gain. This is when the reservoir is in the same state as the state of the apparatus that we are trying to erase. In this case we just have a state swap between the new pure state of the apparatus and our old state ϱ. Curiously enough, creating correlations is not costly in terms of the entropy of the environment (such as when Alice and Bob communicate).

Landauer's principle is a statement that is equivalent to the Second Law of Thermodynamics. If we could delete information without increasing entropy, then we could construct a machine that completely converts heat into work with no other effect which contradicts the Second Law. The converse is also true. Namely if we could convert heat into work with no other effect, then we could use this energy to delete information with no entropy increase (Penrose 1973). Thus, the relative entropy provides an interesting link between thermodynamics, information theory, and quantum mechanics (see also Brillouin (1956)).

6.7 Landauer's erasure and data compression

I shall now show how Landauer's principle can be used to derive the limit to quantum data compression. The entropy lost in deleting the information stored in a string of n qubits all in the state ϱ is $nS(\varrho)$. However, we could first compress this string and then delete the resulting information. The entropy loss after the compression is $m \log 2 = m\beta^{-1}$, where the string has been compressed to m qubits. The entropies before and after compression should be equal if no information is lost during compression, that is, if we wish to have maximal efficiency, and therefore $m/n = S(\varrho)$, as shown previously. Equality is, of course, only achieved asymptotically, as we have argued many times before.

6.8 Summary

The key result in this chapter is the Holevo bound. It expresses an upper bound (asymptotically achievable) on our ability to discriminate between mixed states ϱ_i, prepared with respective probabilities p_i. The expression for the Holevo bound is

$$\xi = S(\varrho) - \sum_i p_i S(\varrho_i) \tag{6.31}$$

where $\varrho = \sum_i p_i \varrho_i$.

We have also argued that in order to delete a certain amount of information stored in a system, we have to increase the entropy of its environment by at least as much as the entropy quantifying that information. This law is called Landauer's erasure principle. It is equivalent to the Second Law of Thermodynamics and provides a nice insight into the various entropic results that we have been deriving rigorously using the basic principles of quantum mechanics.

Part II

Quantum Entanglement

7
Quantum entanglement—introduction

In Chapter 2, we discussed the Mach–Zehnder interferometer experiment in order to show why and how quantum mechanics is different from classical mechanics. A photon sent through a beam splitter behaves like a particle when it is observed by only one of the two detectors. When two beam splitters are used, the photon "interferes with itself" (this is a very vague and archaic phrase—but there just isn't a better shorter one) and behaves like a wave. This is the so called wave–particle duality of quantum mechanics which leads to entanglement.[1]

Here we would like to talk about superpositions when we have two or more particles. This leads to the phenomenon of entanglement which we touched upon before, but we shall now analyze it in full detail. Understanding and analyzing entanglement is one of the most interesting directions in the field of quantum information.

7.1 The historical background of entanglement

Entanglement has puzzled physicists for many years, almost since the birth of quantum mechanics. In 1924, Heisenberg developed the first coherent mathematical formalism for quantum theory. In 1927, he showed that there was a complementarity between classical concepts in quantum mechanics: it was impossible to measure both the position and the momentum of a particle—the Heisenberg uncertainty principle. Contrary to this principle, it is possible to measure the momentum and position of a classical particle. Heisenberg therefore assumed that there is some cutoff point between the classical and quantum worlds. Various other people have then extended and applied Heisenberg's ideas: Schrödinger, Born, Jordan, and Dirac have made the most significant contributions.

In 1935, Einstein, Podolsky, and Rosen (EPR) developed a thought experiment called the "EPR paradox" to demonstrate what they felt was a lack of completeness in quantum mechanics. A lack of completeness would mean that there are some things

---

[1] I should also say that the "wave–particle duality" picture is also very old-fashioned. According to our most advanced understanding of physics—quantum field theory—the fundamental entities in this world are fields, whose vibrations are, in fact, particles. So an electron is just a vibration (excitation) of the electron field, a photon is just an excitation of the electromagnetic field, and Vlatko Vedral is just an excitation of the Vlatko Vedral field (this is not a fundamental field, but is made up of many electron, quark and photon fields). Entanglement can be treated very naturally within field theory, as will be seen at the end of this chapter.

out there in the real world that the quantum formalism is unable to describe. This would be very bad news indeed for quantum mechanics.

The logic of Einstein, Podolsky, and Rosen was as follows. If two entangled particles are in the state

$$|\Phi^+\rangle = \frac{1}{\sqrt{2}}(|0_A 0_B\rangle + |1_A 1_B\rangle) \tag{7.1}$$

(known as an EPR pair), where they could be light years apart, then a measurement made on the first particle seems to affect a measurement on the second particle—if Alice measures the first particle as a $|0\rangle$ then so will Bob, and similarly for $|1\rangle$. Einstein had previously developed the theory of relativity, which says that nothing can travel faster than the speed of light. He called the quantum "instantaneous collapse" effect "spooky action at a distance" and claimed that, if we do not abandon relativity, it leads us to conclude that the quantum description of the two systems is incomplete. The reason was that we can seemingly determine complementary properties of a particle (the position and momentum for example) at the same time, by using information from its entangled partner. This would be contrary to Heisenberg's uncertainty principle, which is a cornerstone of quantum mechanics. I shall not try to go into the details of this argument since it has been debated a huge number of times, both in the technical and in the popular literature, and it is more of historical value rather than being useful for the present understanding of entanglement.[2]

The "hidden-variable" hypothesis—namely that we should be able to go beyond the uncertainty principle—which followed from incompleteness remained controversial for many years until 1964. That was when Bell proposed an experimental test which would be able to invalidate a local hidden-variable hypothesis (Bell 1987).[3] He found various inequalities which a probability distribution would satisfy and showed that two entangled quantum systems would violate these inequalities. (Nevertheless, there are still "hidden-variable" hypotheses which remain today—but they are global in character, i.e. they allow nonlocality. The best-known of these is Bohm's, developed in 1952, which says that the interaction of particles is governed by nonlocal field lines. This interpretation does not contradict Bell's inequalities because a nonlocal element is allowed.)

Round about the time of the EPR paradox, Schrödinger wrote a couple of very influential papers where he pointed out that entanglement is the defining feature of quantum mechanics, but that it could lead to some very counterintuitive results in the macroscopic world. Schrödinger pointed out that macroscopic objects (such as a cat) can become entangled with microscopic objects (such as an atom). This could lead to states where an atom has decayed and kills a cat or the atom has not decayed and the cat is alive. The trouble is that according to quantum mechanics, both possibilities would be present simultaneously! So the cat would be dead and alive at the same time, as it were. This apparent paradox is known as the Schrödinger cat paradox and still generates a great deal of lively debate.

---

[2] Some people would disagree that this argument has only historical value and would argue that the EPR paradox has never been resolved properly (especially not by Bohr!).

[3] Global hidden-variables cannot be eliminated—we could even think of quantum mechanics as a global hidden-variable theory!

For completeness, I shall mention two more developments relevant to entanglement and the foundations of quantum mechanics. The first is a view of quantum mechanics motivated by von Neumann's work in 1930 which showed that entanglement can be used to explain measurements without referring to probability theory (we shall talk about this more later). His work was also foreshadowed by Mott and Heisenberg. The standard "Copenhagen" interpretation of quantum mechanics says that a measurement collapses a state into the basis of measurement with the appropriate probabilities. Von Neumann showed that measurement collapse could be "explained" by the entanglement of the measurement apparatus with the system being measured. After measurement, the apparatus and the system are in an entangled superposition of all possible outcomes. In 1957, motivated by von Neumann's work, Everett published the many-worlds interpretation of quantum mechanics, which says that when a measurement is made, the universe "splits" into a superposition of all the possible measurement outcomes, each of which is an equally real universe in its own right.[4] Therefore, according to this view, the entire universe is governed by unitary dynamics and there is never any collapse and any randomness is excluded from the picture. The third postulate of quantum mechanics (which tells us what happens when we measure) would therefore become superfluous.

7.2 Bell's inequalities

I shall now review Bell's inequalities, but more in the spirit of being able to use them to qualify and quantify entanglement than in the EPR spirit of trying to find paradoxes in quantum mechanics (there are no paradoxes in quantum mechanics, really).

Imagine the following experiment—this is more or less what Bell did. Alice and Bob have a particle from a pair of particles, which they have created with a quantum experiment (the details of this are unimportant). Alice has particle A and Bob has particle B. Alice and Bob each have two sets of measuring apparatus and they agree that they will each, simultaneously and independently, choose one measurement apparatus and then make a measurement with that apparatus. Alice calls her measurement apparatuses A_1 and A_2, and Bob calls his B_1 and B_2. For convenience, we say that each measurement has two possible outcomes $+1$ and -1 (the values of the outcome are arbitrary and can be changed). Bell's inequalities then state that for a local hidden-variable theory:

$$E(A_1 B_1) + E(A_1 B_2) + E(A_2 B_1) - E(A_2 B_2) \leq 2 \tag{7.2}$$

where $E(A_i B_j)$ is the expectation value of when Alice measures with apparatus A_i and Bob measures with apparatus B_j. Bell made two key assumptions:

1. Each measurement reveals an objective physical property of the system. This means that the particle had some value of this property before the measurement was made, just as in classical physics. This value may be unknown to us (just as it is in statistical mechanics), but it is certainly there.

---

[4]For every wrong choice you have made in this universe, there is a parallel universe where you have made all the right choices. In some other universe I have a good sense of humor, and am very good-looking, smart and rich. Also, Schrödinger's cat really is dead and alive, but in different universes!

2. A measurement made by Alice has no effect on a measurement made by Bob and vice versa. This comes from the theory of relativity, which requires that any signal has to propagate at the (finite) speed of light.

To prove the inequality, we write the sum of the expectation values as

$$E(A_1 B_1) + E(A_1 B_2) + E(A_2 B_1) - E(A_2 B_2) \tag{7.3}$$
$$= E(A_1 B_1 + A_1 B_2 + A_2 B_1 - A_2 B_2) \tag{7.4}$$
$$= E(A_1(B_1 + B_2) + A_2(B_1 - B_2)) . \tag{7.5}$$

The outcome of each experiment is ± 1, which leads to two cases:
- $B_1 = B_2$. In this case $B_1 - B_2 = 0$ and $B_1 + B_2 = \pm 2$, so $A_1(B_1 + B_2) + A_2(B_1 - B_2) = \pm 2 A_1 = \pm 2$.
- $B_1 = -B_2$. In this case $B_1 + B_2 = 0$ and $B_1 - B_2 = \pm 2$, so $A_1(B_1 + B_2) + A_2(B_1 - B_2) = \pm 2 A_2 = \pm 2$.

So in either case, $A_1 B_1 + A_1 B_2 + A_2 b_1 - A_2 B_2 = \pm 2$. We therefore obtain the following Bell's inequality:

$$E(A_1 B_1) + E(A_1 B_2) + E(A_2 B_1) - E(A_2 B_2) \tag{7.6}$$
$$= E(A_1 B_1 + A_1 B_2 + A_2 B_1 - A_2 B_2) \tag{7.7}$$
$$= \sum_{a_1, a_2, b_1, b_2} p(a_1, a_2, b_1, b_2)(a_1 b_1 + a_1 b_2 + a_2 b_1 - a_2 b_2) \tag{7.8}$$
$$\leq 2 . \tag{7.9}$$

Quantum mechanics can violate this inequality, and this has been observed experimentally with many systems over last 30 years or so. Let's see how Alice and Bob can violate Bell's inequalities. For that, we need to choose two observables for Alice and two observables for Bob. Different choices will lead to different numbers, and we are looking for the choice that achieves the maximum. We also know that, classically, we cannot exceed the number 2. It turns out that the best choice is that Alice measures σ_3 and $\sigma_3 + \sigma_1$ and Bob measures $(1/4)\sigma_3 + (3/4)\sigma_1$ and $(3/4)\sigma_3 + (1/4)\sigma_1$. (These correspond to angles of measurement of 0, 45, 22.5 and 67.5 degrees, respectively). The inequality with this choice is

$$E(A_1 B_1) + E(A_1 B_2) + E(A_2 B_1) - E(A_2 B_2) \tag{7.10}$$
$$= \frac{3\sqrt{2} + (-1)(-\sqrt{2})}{2} = 2\sqrt{2} , \tag{7.11}$$

and this is well above 2! An EPR pair therefore violates Bell's inequalities, and this can be shown to be the maximal total violation for two qubits.

So, what was wrong with the argument leading up to the bound of 2?[5] Various physicists—depending on their prejudices and personalities—would now conclude that quantum mechanics violates Bell's assumption 1 or Bell's assumption 2 (or both!).

---

[5] Mathematically, the mistake was, speaking somewhat imprecisely, to assume that the expectation value of a product of observables was the product of the expectation values, which is clearly not the case when states are entangled.

Another possibility is to assume the Everett–von Neumann picture of measurements, which avoids probabilities altogether, and remains on the level of probability amplitudes all the time. In that case Bell's inequalities make no sense to start with, and this is certainly a way out of "quantum paradoxes"! We are not interested here in the philosophical implications of all this,[6] but we are interested in whether all entangled states violate Bell's inequalities and whether states that are more entangled violate them by a higher amount. An excellent account of all such philosophical issues can be found in the collection of Bell's papers in Bell (1987).

7.3 Separable states

First, we introduce a large family of states that definitely do not violate Bell's inequalities. They are called "separable" or "disentangled" states in the literature. A separable state is (by definition) a state of the form[7]

$$\varrho_{AB} = \sum_i p_i \varrho_A^i \otimes \varrho_B^i . \tag{7.12}$$

We can easily see that this state cannot violate Bell's inequalities, since for each i, the joint state of Alice and Bob is just a product state $\varrho_A^i \otimes \varrho_B^i$. Measurements by Alice and Bob therefore behave completely independently and this is not changed when probabilities are introduced on top. The reader can verify that indeed

$$E(AB) = \mathrm{tr}(A \otimes B \varrho_A^i \otimes \varrho_B^i) = \mathrm{tr}(A\varrho_A^i)\mathrm{tr}(B\varrho_B^i) \tag{7.13}$$

from which it follows that the combination of operators in Bell's inequalities can never exceed a value of 2.

Another reason for calling these states "disentangled" is that a separable state can always be prepared under conditions of local operations and classical communication (LOCC). LOCC means that Alice and Bob communicate classically, and perform quantum operations only on their own systems (these could be any CP-map operations, but restricted to their own systems). Mathematically this means that what Alice and Bob are restricted to operations of the form $A \otimes B$, and then exchanging classical communication. Any sequence of these is allowed, but nothing more general is.

Separable states are, in fact, the most general class of states that can be prepared by LOCC. There is a very simple way of doing this, which is for Alice to prepare a state ϱ_A^i with some probability p_i and then phone Bob up (this is the classical-communication part) and inform him that he should prepare the state ϱ_B^i.

Before we proceed with analyzing mixed-state entanglement, we shall first show that all pure states that are entangled violate Bell's inequalities. Therefore, Bell's inequalities are an excellent criterion for pure-state entanglement. But there are many other ways of characterizing two-system pure-state entanglement.

---

[6]I was once asked by a colleague of mine how high I ranked Bell's achievement on a scale of 1 to 100 (where 1 is the lowest and 100 is the highest ranking). My answer was 2, which was a big surprise to him (I suppose he ranked it close to 100). I believe that Bell's inequalities are very important historically, but I actually view rather negatively most (but not all) of the philosophical debate that has surrounded them ever since.

[7]First introduced by Werner in 1989.

7.4 Pure states and Bell's inequalities

Let $|\psi^{AB}\rangle$ be a pure state of two entangled qubits. Using the Schmidt decomposition, $|\psi^{AB}\rangle$ can be written as

$$|\psi^{AB}\rangle = \alpha|00\rangle + \beta|11\rangle \qquad (7.14)$$

where the $|0\rangle$'s and $|1\rangle$'s are orthonormal basis states. We can write $|\psi^{AB}\rangle$ as

$$|\psi^{AB}\rangle = \alpha(|00\rangle + |11\rangle) + (\beta - \alpha)|11\rangle . \qquad (7.15)$$

Therefore, intuitively, $|\psi^{AB}\rangle$ contains part of the maximally entangled Bell state which can be used to violate Bell's inequalities. This can be proven formally by calculating the expectation values of Alice's and Bob's operators in the same way as before.[8] However, there is a more instructive way of showing the violation of Bell's inequalities.[9]

We can show that $|\psi^{AB}\rangle$ violates Bell's inequalities by performing a local CP-map which takes $|\psi^{AB}\rangle$ into a mixture of $|\Phi^+\rangle$ and $|01\rangle$. Since this operation is local, it cannot increase the entanglement (a observation that will be used extensively in the next chapter), and therefore if the final state is found to violate Bell's inequalities, the original state must also be entangled.

Let U^A be a unitary operation defined by

$$U^A|0\rangle|0\rangle = |0\rangle|0\rangle , \qquad (7.16)$$
$$U^A|0\rangle|1\rangle = \alpha|0\rangle|1\rangle + \beta|1\rangle|0\rangle . \qquad (7.17)$$

If Alice has an auxiliary qubit, we can write the initial system as

$$|0^A\rangle|\psi^{AB}\rangle . \qquad (7.18)$$

When Alice applies the unitary operation locally to her qubits, we obtain

$$U^A \otimes I^B(|0^A\rangle|\psi^{AB}\rangle) = a|000\rangle + b(\alpha|011\rangle + \beta|101\rangle) \qquad (7.19)$$
$$= |0\rangle(a|00\rangle + b\alpha|11\rangle) + b\beta|101\rangle . \qquad (7.20)$$

Therefore, if we tailor the unitary transformation so that $a = b\alpha$, then if Alice measures her ancillary qubit in the state $|0\rangle$, the state that she shares with Bob is maximally entangled. And this maximally entangled state violates Bell's inequalities. So what we have shown is that by a local unitary transformation followed by a measurement, Alice can convert any nonmaximally entangled pure state into a maximally entangled pure state (with some nonzero probability). Since we get a Bell violation after this procedure we must conclude that the initial state contained some entanglement to start with. The procedure of amplifying entanglement by local operations will be key to quantifying entanglement in Chapter 9, when we are going to analyze the quantifying of entanglement in much more detail.

---

[8]This was done by Gisin in 1991.
[9]This was done by Gisin again in 1995.

7.5 Mixed states and Bell's inequalities

Although all pure entangled states violate Bell's inequalities, this not true for entangled mixed states, which makes it much harder to qualify (and quantify) mixed state entanglement in general. To illustrate this, we shall look at a particularly simple class of (one-parameter) mixed states, known as Werner states.

The Werner states are defined as mixtures of Bell states, where the degree of mixing is determined by a parameter F (which really stands for "fidelity"):

$$\varrho_W = F|\Psi^-\rangle\langle\Psi^-| + \frac{1-F}{3}(|\Psi^+\rangle\langle\Psi^+| + |\Phi^+\rangle\langle\Phi^+| + |\Phi^-\rangle\langle\Phi^-|), \qquad (7.21)$$

where $0 \leq F \leq 1$. When $F = 1/2$, we can write ϱ_W as

$$\begin{aligned}\varrho_W &= \frac{1}{6}(|\Psi^-\rangle\langle\Psi^-| + |\Psi^+\rangle\langle\Psi^+|) + \frac{1}{6}(|\Psi^-\rangle\langle\Psi^-| + |\Phi^+\rangle\langle\Phi^+|) \\ &+ \frac{1}{6}(|\Psi^-\rangle\langle\Psi^-| + |\Phi^-\rangle\langle\Phi^-|)\end{aligned} \qquad (7.22)$$

This is a fully separable state, simply because an equal mixture of any two maximally entangled states is a separable state. For example, $(1/2)(|\Phi^+\rangle\langle\Phi^+| + |\Phi^-\rangle\langle\Phi^-|)$ can be written as $(1/2)(|00\rangle\langle00| + |11\rangle\langle11|)$, which is clearly separable. When $F > 1/2$, ϱ_W cannot be expanded out in this form and is entangled. We shall show that the Werner states are entangled for $F > 1/2$ in the following chapter. However, ϱ_W only violates Bell's inequalities when $F > 0.78$.[10] Therefore Bell's inequalities are not a necessary and sufficient indicator of entanglement.

Since Bell's inequalities are in general insufficient for discriminating mixed entangled states from separable (disentangled) states, we need to design a better criterion for this purpose. In the following two chapters we shall talk about Hermitian operators—known as entanglement witnesses—which can be constructed for discrimination of entangled and disentangled states. After we have explained in detail how to qualify entanglement—this is of course still a topic of intense research activity and many open problems remain—we shall move onto the topic of quantifying entanglement.

One very important message from Bell's inequalities is that it is not enough to have correlations in one basis between two systems to have entanglement. We must find another basis which is correlated at the same time. So entanglement really is an excess of correlations allowed by quantum mechanics that has no counterpart in classical physics.

7.6 Entanglement in second quantization

We usually think of entanglement as existing between different degrees of freedom of two or more particles.[11] There have been a number of important advances in understanding entanglement in systems containing a small number of particles. If we are to use this concept appropriately, however, we need to extend our analysis to realistic systems containing a large number of particles. When it comes to large systems

---

[10] I state this without proof, as I intend the discussion only to provide a general motivation for understanding mixed states. The proof here would have no value here other than historical.

[11] This section can be omitted on first reading, and the rest of the book is independent of it.

in quantum mechanics, the concept of a particle fades away and is replaced by the notion of "an excitation of a given mode of the field representing the particle". Individual particles actually become indistinguishable (this is, of course, also true when we have a small number of identical particles, but indistinguishability may not play any significant role). In addition, the concepts of particle statistics (fermions versus bosons) also become directly relevant, and it is important to understand the relation to entanglement. The most convenient and appropriate formalism to deal with all the issues involving a large number of systems is the second quantization. So we need to understand what entanglement means in this setting if we are to be able to harness solid-state and condensed-matter systems for information-processing purposes. Another benefit of second quantization is that it is also the correct formalism for (relativistic) quantum field theory.

We shall see that entanglement is highly dependent on the choice of modes. The choice of modes is mainly dictated by the physics of the given situation, and the general rule is that in order for modes to be entangled they need to interact in some way. The crux is that what is an interacting Hamiltonian for one set of modes may not be so for a different set of modes. Alternatively stated, the vacuum state of one set of modes may not be the vacuum state when the modes are changed. It is this simple fact that will be briefly discussed here. Most interestingly we shall see that the modes need not interact in order to produce entanglement—entanglement can be a consequence of indistinguishability of quantum particles and the existence of quantum statistics.[12]

First we briefly review the formalism of second quantization. We shall approach the second-quantization formalism from the perspective of many-particle systems, rather than as a means of achieving a relativistic quantum theory. Suppose we have the Schrödinger equation for a single quantum system in one dimension (for simplicity):

$$\hat{H}(x)\Psi(x,t) = i\hbar \frac{\partial}{\partial t}\Psi(x,t) \ . \tag{7.23}$$

Note that although this will lead to nonrelativistic second quantization, much of what we say will be true relativistically as well. A formal solution of this equation is

$$\Psi(x,t) = \sum_n b_n(t)\psi_n(x) \ . \tag{7.24}$$

As usual, by substituting this back into the Schrödinger equation, we obtain its time-independent version

$$\hat{H}(x)\psi_n(x) = E_n\psi_n(x) \ , \tag{7.25}$$

where the E_n are the corresponding energies. In order to convert this into a many-particle equation, we apply the formalism of second quantization, which means that we effectively have to "upgrade" the wavefunction Ψ into an operator.[13] Formally, we write

---

[12]Entanglement between modes is entanglement in the numbers of excitations in each of the modes. So, the key question to ask is, if mode 1 contains five quanta, say, what is the probability that mode 2 also contains the same number?

[13]Hence the name "second quantization".

Entanglement in second quantization

$$\hat{\Psi}(x,t) = \sum_n \hat{b}_n(t)\psi_n(x) . \tag{7.26}$$

The operators $\hat{b}_n$ are the well-known annihilation operators.[14] The conjugate of the above equation becomes

$$\hat{\Psi}^\dagger(x,t) = \sum_n \hat{b}_n^\dagger(t)\psi_n^*(x) \tag{7.27}$$

where the $b_n^\dagger$ are the creation operators. Here we consider only time-independent creation and annihilation field operators. Nothing much would change conceptually if the field operators were to be time-dependent, although mathematically the whole analysis would become much more involved. The Hamiltonian also has to be second-quantized, and the new Hamiltonian is given by the average of the old (first-quantized) Hamiltonian

$$\tilde{H} = \int dx \hat{\Psi}^\dagger(x,t) \hat{H}(x) \hat{\Psi}(x,t) \tag{7.28}$$

Where we invoke the orthogonality rules $\langle \psi_n | \psi_m \rangle = \delta_{mn}$, the second-quantized Hamiltonian becomes

$$\tilde{H} = \sum_n E_n \hat{b}_n^\dagger \hat{b}_n , \tag{7.29}$$

and this is the same as that of a set of independent harmonic oscillators. Now, the fields here are clearly noninteracting and therefore we may conclude that the b modes are disentangled. But, we have to be careful. While there is, clearly, no entanglement in the b modes, there may be entanglement between some other modes. To illustrate, this let us have a look at the simplest case of two harmonic oscillators. Note that the index n can also contain internal degrees of freedom (such as spin or polarization).

Suppose that we have two noninteracting harmonic oscillators so that the total Hamiltonian is the following (we shall omit hats from now on):

$$H = \hbar(\omega - \lambda)b_1^\dagger b_1 + \hbar(\omega + \lambda)b_2^\dagger b_2 . \tag{7.30}$$

Since the oscillators are noninteracting, their eigenstates are composed of disentangled direct products of the eigenstates of the individual harmonic oscillators. For example, the ground state is just the product of the ground states of oscillator 1 and oscillator 2, $|0\rangle_b = |0\rangle_{b_1} \otimes |0\rangle_{b_2}$, so that it is annihilated by both of the annihilation operators of the individual oscillators:

$$b_1 |0\rangle_b = 0 , \tag{7.31}$$
$$b_2 |0\rangle_b = 0 . \tag{7.32}$$

However, let us make a change of modes and see what happens to the new eigenstates. Let the new creation and annihilation operators be

$$a_1 = \frac{b_1 + b_2^\dagger}{\sqrt{2}} , \tag{7.33}$$

$$a_2 = \frac{b_1^\dagger - b_2}{\sqrt{2}} . \tag{7.34}$$

---

[14] Creation and annihilation operators are designed to respectively create and destroy particles with a given energy. The energy levels (values) here carry the label n.

(We could have made a more general change, but the one above is sufficient to illustrate the point here.) The original Hamiltonian now becomes

$$H = \hbar\omega(a_1^\dagger a_1 + a_2^\dagger a_2) + \hbar\lambda(a_1^\dagger a_2 + a_2^\dagger a_1) \tag{7.35}$$

and it is apparent that the a modes actually represent two coupled harmonic oscillators. It is, therefore, not surprising that the ground state of this system is actually entangled when written in the basis of the original harmonic oscillators. This ground state is given by the following (up to a normalization which is not relevant for our discussion):

$$|0\rangle_a \propto \sum_n \left(-i\tanh\frac{4\lambda}{\omega}\right)^n |n\rangle_{b_1} \otimes |n\rangle_{b_2}, \tag{7.36}$$

where $|n\rangle_{b_i}$ is the state containing n quanta in the oscillator b_i. This state is clearly entangled with respect to the eigenstates of the b's, and it is written in the Schmidt decomposition form making it easy to detect entanglement.

Let us now generalize our consideration to an arbitrary field, which can be written as a superposition of the creation and annihilation operators for the various modes:

$$\psi(x) = \sum_n (\psi_n(x) a_n + \psi_n^*(x) a_n^\dagger). \tag{7.37}$$

Here we have, basically, infinitely many harmonic oscillators to sum over. The vacuum state for the a operators is defined by $a_n|0\rangle = 0$ for all n. Now, suppose that we expand the field in terms of a different set of creation and annihilation operators, as follows:

$$\psi(x) = \sum_n (\tilde{\psi}_n(x) b_n + \tilde{\psi}_n^*(x), b_n^\dagger). \tag{7.38}$$

The new operators can always be expanded as a combination of the old operators as

$$b_m = \sum_n \alpha_{mn} a_n + \beta_{mn} a_n^\dagger \tag{7.39}$$

$$a_m = \sum_n \alpha_{mn}^* b_n - \beta_{mn}^* b_n^\dagger \tag{7.40}$$

(the new mode functions $\tilde{\psi}$ can also be expressed in terms of the old ones, but this is not important for our discussion). Owing to the unitarity requirement, the coefficients of transformation have to possess the following properties:

$$\sum_k (\alpha_{ik}\alpha_{jk}^* - \beta_{ik}\beta_{jk}^*) = \delta_{ij} \tag{7.41}$$

$$\sum_k (\alpha_{ik}\beta_{jk} - \beta_{ik}\alpha_{jk}) = 0. \tag{7.42}$$

Now, the point of our argument here is that the new vacuum state, defined by $b_m|\tilde{0}\rangle = 0$ (for all m), will not in general be the same as the old vacuum state (i.e. it will not be annihilated by the old annihilation operators, a_n). The new vacuum state will, for example, have the following number of particles

$$\langle \tilde{0}|a_i^\dagger a_i|\tilde{0}\rangle = \sum_j |\beta_{ij}|^2 \qquad (7.43)$$

with respect to the ith mode of a (and this means that the vacuum of the b modes contains some particles with respect to the a modes). Therefore, since the vacuum state $|\tilde{0}\rangle$ is an overall pure state, we can conclude that the amount of entanglement between the ith mode and all the other modes taken together can be quantified by, for example, the von Neumann entropy[15]

$$E_i = -\sum_j |\beta_{ij}|^2 \ln |\beta_{ij}|^2 \, . \qquad (7.44)$$

We see that this entanglement is zero only when one of the β's is unity and the rest are zero, that is, the new modes are the same as the old ones. The maximum entanglement, on the other hand, is achieved when all the β's are equal.

In summary, there are two main messages in this section: one is that entanglement is highly dependent on the way we decide to divide the whole system into different modes, and the other is that interaction in general gives rise to entanglement. There are numerous examples of various kinds of mode transformations in quantum optics (for example, in two-mode squeezing), condensed-matter physics, and relativistic quantum field theory which lead to the generation of entanglement as described above.

We shall now forget modes and second quantization and return to describing entanglement using particles in first quantization. Bear in mind, however, that everything we shall say can also be translated to quantum fields.

7.7 Summary

Separable states are defined here to be the most general disentangled states. They are of the form

$$\varrho^{AB} = \sum_i p_i \varrho_i^A \otimes \varrho_i^B \qquad (7.45)$$

where A and B are the labels of two subsystems.

All separable states satisfy Bell's inequalities, however, not all entangled states violate them. Maximally entangled states violate Bell's inequalities, and so do all pure entangled states.

Entangled states can be manipulated locally, and we can create a more entangled state out of a less entangled state by only local operations aided by classical communication (referred to as LOCC).

Ultimately, in quantum field theory, all particles are just excitations of some field. Different modes give rise to particles (say photons) with different properties (say frequency). Here, the entanglement is no longer between internal degrees of freedom of particles (these define different modes as well), such as in a spin-correlated singlet state of two electrons, but it is between the numbers of particles in different modes.

---

[15] We shall see in Chapter 10 that this is, in some sense, a unique measure of entanglement for pure states.

8
Witnessing quantum entanglement

We have seen that entanglement is a key resource in quantum information. Teleportation, dense coding, and many other quantum protocols rely on the existence of entanglement, and would not be possible with just classical correlations. Furthermore, using entanglement between qubits that support computation, quantum computers can solve problems faster than classical computers (as we shall see in the last part of the book, though I must admit that a formal proof of a link between computational speedup and entanglement does not really exist[1]). We have also seen that Bell's inequalities can sometimes help us discriminate between entangled and separable states. However, these inequalities are not always reliable—this is true in the sense that they may fail to detect genuinely entangled states—and here we would like to talk about the most general way of addressing this question. We shall talk all the time using bipartite systems, but all of what will be said will be applicable to the most general multipartite systems.

The aim of entanglement detection is the following: given a bipartite system ϱ^{AB}, we have to decide whether ϱ^{AB} is entangled or separable. Classically, witnessing correlations is pretty trivial. Given a joint state XY, there are correlations between X and Y if (and only if) the mutual information $I(X:Y)$ is positive (this can even be generalized to a larger number of random variables, but one has to be much more careful—a full discussion of how to address classical correlations is well beyond the scope of this book[2]). For pure quantum states, we can look at the Schmidt decomposition or trace out one of the two systems to see if there is entanglement. Entanglement verification for mixed states is very different. Detecting entanglement in mixed states means detecting quantum correlations whilst ignoring classical correlations.[3] And this is the main difficulty: how do we subtract the classical correlations from the total to be left with the entanglement only? Several ways have been proposed for detecting entanglement in mixed states, each with their various advantages and disadvantages.

---

[1]There are two arguments why there is a connection between the two. One is that the set of disentangled states is much smaller than that of entangled states (this can be phrased very rigorously mathematically, but I am not very concerned about the point here). The second—and not entirely unrelated—argument is that disentangled states can be efficiently simulated on a classical computer.

[2]The basic issue is how to define the mutual information for three and more random variables, and this requires a bit more thinking, but nothing as intricate as quantifying entanglement.

[3]For a pure state this issue is simple because classical and quantum correlations are inseparable! Namely, the existence of classical correlations implies entanglement and vice versa.

8.1 Entanglement witnesses

An entanglement witness W is a Hermitian operator which helps us to decide whether a state is entangled or not. The basic idea is that the expectation value of the witness will be different for separable and entangled states.[4] Let me explain how this works in more detail.

We are going to borrow some very simple geometric ideas to explain the construction of W. Let $\mathcal{T}$ be the set of all density matrices (Hermitian operators with positive spectrum and unit trace), let $\mathcal{E}$ be the set of entangled states, and let $\mathcal{D}$ the set of all disentangled (separable) states. Both $\mathcal{E}$ and $\mathcal{D}$ are clearly subsets of $\mathcal{T}$.

The set of all disentangled states $\mathcal{D}$ is convex. If ϱ_1^{AB} and σ_2^{AB} are separable states then they can be written in the form

$$\varrho_1^{AB} = \sum_i p_i \varrho_i^A \otimes \varrho_i^B \tag{8.1}$$

$$\sigma_1^{AB} = \sum_i q_i \sigma_i^A \otimes \sigma_i^B \tag{8.2}$$

(where of course ϱ_i^A and σ_i^A can be distinct from ϱ_i^B and σ_i^B respectively). A linear combination of ϱ_1^{AB} and σ_2^{AB} can be written as

$$p\varrho_1^{AB} + (1-p)\sigma_2^{AB} = p\sum_i p_i \varrho_i^A \otimes \varrho_i^B + (1-p)\sum_i q_i \sigma_i^A \otimes \sigma_i^B \tag{8.3}$$

and is also a disentangled state. The convexity of the set of separable states is a key property in witnessing entanglement:

- We can detect entanglement if we can determine whether or not a given state is in the set of disentangled states.
- A corollary of the Hahn–Banach theorem (whose full statement is irrelevant for our purposes here) is that given a convex set and a point outside it, there exists a plane such that the point is on one side of it and the set is on the other side. I shall treat this corollary as "obvious" (for physicists at least) and will not present a proof.[5]

$\mathcal{D}$ is convex, so we can use this corollary to decide whether a state ϱ is entangled. The entanglement witness is exactly the operator defining this plane which separates our given entangled state from the set of separable states.

Let us express these ideas mathematically using some simple linear algebra. Given an arbitrary vector space V, a plane in V can be defined by the vectors $|\psi\rangle$ that satisfy

$$\langle w|\psi\rangle = 0 , \tag{8.4}$$

where w is a unit vector orthogonal to the plane. So a plane is defined by a single vector such that all the vectors in the plane are orthogonal to it.

---

[4]This is a very natural method: entangled and separable states are discriminated according to different experimental results when one and the same quantity is measured (i.e. the same witness operator).

[5]The proof of the Hahn–Banach theorem is apparently one of the most difficult exam questions for undergraduate mathematicians. There are obviously many intricacies here that are completely invisible to an (unperceptive) physicist such as myself.

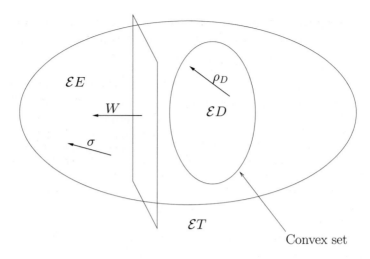

Fig. 8.1 $\mathcal{D}$, the set of disentangled states, is convex. If σ is an entangled state then there exists a Hermitian operator W which defines a plane that lies between $\mathcal{D}$ and σ. W is called an entanglement witness.

We can also think of Hermitian operators as vectors, and we can define an inner product (you can check for yourself that this is an inner product as defined in Chapter 2) as

$$\langle O_1|O_2\rangle = \text{tr}(O_1^\dagger O_2) \,. \tag{8.5}$$

The eigenvalues of a Hermitian operator are real, and so

$$\langle O_1|O_2\rangle = \text{tr}(O_1 O_2) \,. \tag{8.6}$$

We can use this inner product to define a plane in the space of Hermitian operators $\mathcal{T}$,

$$\text{tr}(w\varrho) = 0 \,, \tag{8.7}$$

where W is our entanglement witness (see Fig. 8.1). The zero on the right-hand side is there by convention—any other constant will, of course, do the job.

Suppose σ is an entangled state. There exists a plane (shown in Fig. 8.1) defined by a Hermitian operator W such that σ is on one side and the convex set of all the disentangled states is on the other. The points on the plane are defined by $\text{tr}(\varrho W) = 0$. The trace operator is continuous, so for each disentangled state ϱ_D either $\text{tr}(\varrho_D W) \geq 0$ or $\text{tr}(\varrho_D W) \leq 0$. We pick W so that $\text{tr}(\varrho_D W) \leq 0$ and then, for all the points on the other side of the plane which are entangled, $\text{tr}(\sigma W) > 0$. A state σ is entangled if and only there is a (Hermitian) operator W such that $\text{tr}(W\sigma) < 0$, whereas for all separable states ϱ_D we have that $\text{tr}(W\varrho_D) \geq 0$. This is our test for entanglement.

The use of entanglement witnesses is mathematically elegant but practically implausible. As the dimensionality of the system becomes large, the entanglement witnesses can become very hard to find (since there is a continuous infinity of them to search over!). The good news is that if a state is entangled, a witness always exists, but the bad news is that finding it is very difficult in general. It is only under very special circumstances—such as when we have two qubits—that we can obtain an altogether simpler criterion for entanglement. We turn to this topic now.

8.2 The Jamiolkowski isomorphism

It turns out that entanglement witnesses can be expressed in terms of positive maps, rather than in terms of Hermitian operators. This can sometimes be of advantage as we shall see shortly.

Positive maps are known as superoperators, because their purpose is to act on operators and produce other operators. For example, the dynamics of a quantum system that produces a density matrix (the final state) out of another density matrix (the initial state) is a superoperator. We have seen that all physical evolution within quantum mechanics can be described by completely positive, trace-preserving superoperators (Chapter 3). So why are we now giving up the "completely" when constructing entanglement witnesses?

Before we go further, let us first introduce some useful definitions:

- A map is linear if $\Phi(\alpha M + \beta N) = \alpha \Phi M + \beta \Phi N$.
- A map is self-adjoint if it maps Hermitian operators onto Hermitian operators: $\epsilon(M) = \epsilon^\dagger(M) \forall M = M^\dagger$.
- A map is positive if it is linear and self-adjoint and maps positive operators onto positive operators.
- A map ϵ is completely positive if it is positive and if any extension of the form $\tilde{\epsilon} : B(H_A \otimes H_B) \to B(H_B \otimes H_C) \tilde{\epsilon} = I \otimes \epsilon$ for any dimension of A is also positive in the extended operator space.

Therefore, complete positivity requires that the result of the application of a map (i.e. a superoperator) is in fact positive not only in the domain of its operation, but also in any extension. This is a very natural physical requirement for CP-maps. If I perform a physical operation in my room, then I would like that operation also to be physical on the cosmic level (i.e. nothing funny should happen elsewhere in the universe as a result of my local actions in my vicinity). But if a map is positive (but not a CP-map), then by acting on a part of the system, it could cause the whole system to do something unphysical. And this is the whole point of witnessing entanglement with positive maps (but not CP-maps): if we apply a positive map to one of two entangled subsystems, then the total state might lead to a negative operator (and therefore not an allowed density matrix). To show this we need to switch from entanglement witnesses in the form of Hermitian operators to entanglement witnesses in the form of positive maps.

The Jamiolkowski isomorphism is exactly designed for this purpose, relating as it does Hermitian operators to positive maps. The isomorphism can be written in the form
$$A = I \otimes \Phi_A(|\phi^+\rangle\langle\phi^+|), \qquad (8.8)$$
where A is a Hermitian operator and Φ_A is a positive operator, and
$$|\phi^+\rangle = \frac{1}{\sqrt{N}} \sum_{n=1}^{N} |nn\rangle \qquad (8.9)$$
is the maximally entangled state of two N-dimensional systems (N could be any number).

Proof. This theorem states that any Hermitian operator can be obtained from a maximally entangled state of two systems by acting on one of them with a positive map.

If A is Hermitian, then we can always write it in its eigendecomposition form: $A = \sum_i a_i |a_i\rangle\langle a_i|$, where the a_i's are its eigenvalues and the $|a_i\rangle$'s are the corresponding eigenvectors. Let us suppose for the moment that A is in addition positive, that is, $A \geq 0$, meaning that all the eigenvalues are positive numbers (including zero). We can then define operators

$$I \otimes M_i |\phi^+\rangle = \sqrt{a_i}|a_i\rangle \,, \tag{8.10}$$

whose function is to produce the eigenstates (with the appropriate prefactor) out of $|\Phi^+\rangle$. Clearly, $M_i^\dagger M_i$ is a positive operator and the sum of the action of all operators M therefore results in the positive operator A:

$$A = I \otimes M_i |\phi^+\rangle\langle\phi^+| I \otimes M_i^\dagger \,, \tag{8.11}$$

where $\Phi_A = \sum_i M_i(.)M_i^\dagger$ is completely positive.

It is now clear that if we need A to have negative eigenvalues as well, then Φ_A needs to be only positive, and we should not demand complete positivity. This completes the proof of the isomorphism □.

We can now state a theorem showing that positive maps can be used as entanglement witnesses.

Theorem. A state σ_{12} is entangled if and only if there exists a positive map Λ (not a CP-map) such that

$$I \otimes \Lambda(\sigma_{12}) < 0 \,. \tag{8.12}$$

Proof. Once we have the Jamiolkowski isomorphism in place, it is very easy to prove this theorem from our previous result using Hermitian entanglement witnesses. Previously we said that if the average of the witness for some state is negative (while it is positive for all separable states) then this state is entangled. Now we only need to convert this statement into one dealing with positive maps, and this is done as follows:

$$\text{tr}(W\sigma_{12}) = \text{tr}\{I \otimes \Lambda(|\phi^+\rangle\langle\phi^+|)\sigma_{12}\} \tag{8.13}$$
$$= \langle\phi^+| I \otimes \Lambda(\sigma_{12})|\phi^+\rangle \,, \tag{8.14}$$

where the last line follows from the cyclic property of the trace and the fact that $\text{tr}(O|\alpha\rangle\langle\alpha|) = \langle\alpha|O|\alpha\rangle$. Therefore, if $\text{tr}(W\sigma_{12}) < 0$ then (and only then) $\langle\phi^+| I \otimes \Lambda(\sigma_{12})|\phi^+\rangle < 0$ and so $I \otimes \Lambda(\sigma_{12}) < 0$. And so we have shown the chain of reasoning that leads from the fact that the state is entangled to the fact that there exists a positive map whose action on one subsystem will yield a negative operator in the end .□

This means that if we find a positive map that, when applied to one of the subsystems, leads to a negative operator in the end (a negative operator is one that has negative eigenvalues), the original state must be entangled. Therefore, the search for positive operators and entanglement witnesses is one and the same.

Note that if the state is separable, then any action of a positive map on one of its subsystems still yields an overall positive (density) matrix. This can be seen in the following:

$$I \otimes \Lambda \left(\sum_i p_i \varrho_i^1 \otimes \varrho_i^2 \right) = \sum_i p_i \varrho_i^1 \otimes \Lambda(\varrho_i^2) > 0 , \tag{8.15}$$

which shows that the overall action of a positive map on a separable state always leads to a positive operator in the end.

This remarkable isomorphism between positive maps and Hermitian operators will now be translated into a powerful theorem when applied to two qubits. To present this theorem, we need another mathematical result. The statement is that in two dimensions any positive map P can be written as

$$P = CP_1 + CP_2 T , \tag{8.16}$$

where CP_1 and CP_2 are completely positive maps and T is a transposition operation. This is a very convenient decomposition, which effectively says that the action of a positive map on a two-dimensional system (a qubit system) can yield unphysical results only through the action of the transposition.

Applying this result to entanglement proceeds as follows. Suppose that a positive map is applied to one of the two qubits in some joint state ϱ_{AB}. Then,

$$I_A \otimes P_B(\varrho_{AB}) = I_A \otimes (CP_1 + CP_2 T)_B(\varrho_{AB}) , \tag{8.17}$$

and the last expression is positive (or negative) only if the transposition of B in the state ϱ_{AB} is positive (or negative, respectively). Therefore the state ϱ_{AB} is separable if and only if

$$T_B(\varrho_{AB}) > 0 ; \tag{8.18}$$

that is, its partial transposition is positive. This is very simple to test for in practice and is know as the Peres–Horodecki criterion. The partial transposition is usually denoted by $\varrho_{AB}^{T_B}$.

8.3 The Peres–Horodecki criterion

We have shown that ϱ^{AB} is entangled if and only if there exists an entanglement witness W or if for all positive maps, $I^A \otimes \Phi^B \varrho^{AB} \geq 0$. Given a state ϱ^{AB} it is difficult to show whether or not there exists such an entanglement witness or positive map. The Peres–Horodecki criterion provides a well-known technique which can be used to check whether a state of two qubits is entangled. The positive map used is the transposition operator $\varrho \to \varrho^T$. A state is entangled if $\varrho^{T_B} = I \otimes T \varrho^{AB}$ is positive, where $I \otimes T \varrho^{AB}$ means that we take the transpose of system B, leaving system A as it is.

As an example of the Peres–Horodecki criterion in practice, let us test it on a maximally entangled state $|00\rangle + |11\rangle$, which can be written as a density matrix as follows:

$$\varrho = \frac{1}{2}(|00\rangle\langle 00| + |11\rangle\langle 00| + |00\rangle\langle 11| + |11\rangle\langle 11|) . \tag{8.19}$$

The partial transposition of the second qubit yields

$$\varrho^{T_B} = \frac{1}{2} \left(|0\rangle\langle 0| \otimes (|0\rangle\langle 0|)^T \right.$$
$$+ \left. |1\rangle\langle 0| \otimes (|1\rangle\langle 0|)^T + |0\rangle\langle 1| \otimes (|0\rangle\langle 1|)^T + |1\rangle\langle 1| \otimes (|1\rangle\langle 1|)^T \right) \quad (8.20)$$
$$= \frac{1}{2}(|00\rangle\langle 00| + |10\rangle\langle 01| + |01\rangle\langle 10| + |11\rangle\langle 11|). \quad (8.21)$$
$$(8.22)$$

One eigenvector of ϱ^{T_B} is $|01\rangle - |10\rangle$, which has an eigenvalue $-1/2$. Hence ϱ^{T_B} is not positive, and ϱ^{AB} is entangled. We can now clearly draw the same conclusion for any pure entangled state of two qubits.

We can now apply this to mixed states, which is where Bell's inequalities fail to detect some entangled states. Using the Peres–Horodecki criterion, we can show that Werner states are inseparable. Recall that a Werner state is defined as:

$$\varrho_W = F|\Phi^+\rangle\langle\Phi^+| + \frac{1-F}{3}(|\Phi^-\rangle\langle\Phi^-| + |\Psi^+\rangle\langle\Psi^+| + |\Psi^-\rangle\langle\Psi^-|) \quad (8.23)$$

where $0 \leq F \leq 1$ is known as the singlet fidelity. This state is very important in practice, as it is obtained from the maximally entangled singlet state in the presence of some form of isotropic noise, not uncommon in practice. The partial transposes of the Bell states are

$$|\Phi^+\rangle\langle\Phi^+|^{T_B} = \frac{1}{2}(|00\rangle\langle 00| + |01\rangle\langle 10| + |11\rangle\langle 00| + |11\rangle\langle 11|), \quad (8.24)$$

$$|\Phi^-\rangle\langle\Phi^-|^{T_B} = \frac{1}{2}(|00\rangle\langle 00| - |01\rangle\langle 10| - |10\rangle\langle 01| + |11\rangle\langle 11|), \quad (8.25)$$

$$|\Psi^+\rangle\langle\Psi^+|^{T_B} = \frac{1}{2}(|01\rangle\langle 01| + |00\rangle\langle 11| + |11\rangle\langle 00| + |10\rangle\langle 10|), \quad (8.26)$$

$$|\Psi^-\rangle\langle\Psi^-|^{T_B} = \frac{1}{2}(|01\rangle\langle 01| - |00\rangle\langle 11| - |11\rangle\langle 00| + |10\rangle\langle 10|). \quad (8.27)$$

Hence

$$\frac{1-F}{3}(|\Psi^-\rangle\langle\Psi^-| + |\Psi^+\rangle\langle\Psi^+|)^{T_B} = \frac{1-F}{3}(2|01\rangle\langle 01| + 2|10\rangle\langle 10|),$$

$$\left(\frac{3F}{3}|\Phi^-\rangle\langle\Phi^-| + \frac{1-F}{3}|\Phi^+\rangle\langle\Phi^+|\right)^{T_B} = \frac{1}{3}((2F+1)(|00\rangle\langle 00| + |11\rangle\langle 11|)$$
$$+ (4F-1)(|01\rangle\langle 10| + |10\rangle\langle 01|)). \quad (8.28)$$

The partial transpose of ϱ can be written in matrix form as

$$\varrho^{T_B} = \frac{1}{3}\begin{pmatrix} 2F+1 & 0 & 0 & 0 \\ 0 & 2-2F & 4F-1 & 0 \\ 0 & 4F-1 & 2-2F & 0 \\ 0 & 0 & 0 & 2F+1 \end{pmatrix}, \quad (8.29)$$

and has eigenvalues $2F+1$ for $|00\rangle$, $|11\rangle$, $|01\rangle + |10\rangle$ and $3-6F$ for $|01\rangle - |10\rangle$, and so it is entangled when $F > 1/2$. Therefore, if the singlet fraction is bigger than one-half, the Werner state is always inseparable. Recall also that Bell's inequalities are only violated if $F > 0.78$, and so they are clearly not such a strong entanglement witness.

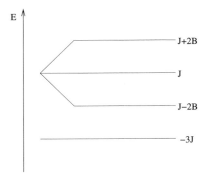

Fig. 8.2 Energy-level diagram of a Heisenberg-coupled dimer of two spins.

8.4 More examples of entanglement witnesses

I would now like to show that there is nothing mysterious about entanglement witnesses and that quite ordinary observables, such as the energy or magnetization of a physical system, can serve as entanglement witnesses.[6] As we have seen entanglement witnesses are operators that "react" differently to entangled states and disentangled states. In particular, the expectation value of a witness is positive for disentangled states and negative for entangled states.

For our purposes, it will be easiest to analyze dimers. Dimers are chains of spins where only pairs of them interact and there is no coupling between the pairs themselves. Therefore they can be described by the following Hamiltonian:

$$H = B(\sigma_z^1 + \sigma_z^2) + J(\sigma_x^1 \otimes \sigma_x^2 + \sigma_y^1 \otimes \sigma_y^2 + \sigma_z^1 \otimes \sigma_z^2), \quad (8.30)$$

where the superscripts are the spin labels and the σ's are the Pauli matrices. The first term is just an external field interacting with the spins, and the second is the Heisenberg coupling between them. The eigenvectors of this system are the singlet state $|\psi^-\rangle = |01\rangle - |10\rangle$ and the triplet states $|00\rangle, |11\rangle$, and $|\psi^+\rangle = |01\rangle + |10\rangle$. The respective eigenvalues are $-3J, J + 2B, J - 2B$, and J (as in Fig. 8.2).

Imagine now that $B = 0$ for simplicity. Let us compute the average value of the Hamiltonian for all separable states. First of all, for all product states,

$$|\langle H \rangle_{\text{prod}}| = J|\langle \sigma^1 \rangle \langle \sigma^2 \rangle| \leq 1, \quad (8.31)$$

where the inequality follows from the fact that the average of the sigma matrices has to lie between -1 and 1. Therefore, the average value of the Hamiltonian for a product state (and hence for any separable state, which is just a convex combination of product states) can never be smaller than $-J$ or greater than J. However, we know

---

[6] We have recently started to realize that macroscopic entanglement is a real possibility and we now have strong evidence that some solids are entangled even at temperatures well above absolute zero. These preliminary investigations have led to an explosion of interest in macroscopic entanglement. We now understand that entanglement can have an observable influence at the macroscopic level and, furthermore, that thermodynamic quantities such as the internal energy and magnetic susceptibility can serve as entanglement witnesses.

that the energy of the singlet state is $-3J$. This state therefore lies outside the domain of separable states. The Hamiltonian is thus found to be an entanglement witness for this system. In fact, any state for which the expectation value of the energy is

$$|\langle H \rangle| > J \tag{8.32}$$

must be entangled. And this is true for mixed states as well.

The thermal state of this system can be written as a Boltzmann mixture of the singlet and the triplet states:

$$\begin{aligned}\varrho_T &= \frac{1}{Z} e^{3\beta J} |\psi^-\rangle\langle\psi^-| + e^{-\beta(J+2B)} |00\rangle\langle 00| \\ &+ e^{-\beta(J-2B)} |11\rangle\langle 11| + e^{-\beta J} |\psi^+\rangle\langle\psi^+|\end{aligned} \tag{8.33}$$

where $Z = e^{3\beta J} + e^{-\beta(J+2B)} + e^{-\beta(J-2B)} + e^{-\beta J}$ is the partition function of the system and $\beta = 1/kT$.

Entanglement in the thermal state of this Hamiltonian has been analyzed extensively, and we shall focus only on the two regimes important for us:

- The first regime is when the coupling J is large compared to the external field B. The ground state is then the singlet state, and at low temperature the system is therefore entangled. At higher temperatures the triplet becomes mixed into the singlet, and when (roughly) $T > J/k$, the entanglement completely disappears (when the external field is zero). Therefore, in order to have high-temperature entanglement in dimers we need a high value of the coupling constant J.
- When J is fixed, the second regime occurs when we can vary the value of the external field B. When B is large (greater than $2J$), the ground state is $|11\rangle$ and at zero temperature the dimers are therefore not entangled. The point where the singlet state becomes replaced by $|11\rangle$ as the ground state ($B = 2J$) is known as the quantum phase transition. However, if we start to increase the temperature, the singlet state—which is the first excited state under these circumstances—starts to become populated. Entanglement can then be generated by increasing the temperature.

Therefore, with entanglement-witnessing techniques, we are now in a position to analyze realistic models of interacting physical systems. Next we address the question of how to measure witnesses in practice.

8.5 Summary

Entanglement witnesses are Hermitian operators which separate disentangled from entangled states. The expectation value of an entanglement witness for an entangled state is different from the value for any separable state.

Finding entanglement witnesses is in general very difficult (because of their multitude). However, they can be rephrased in terms of positive maps. If there is a positive map that acts on one of the subsystems with the result that the overall state is negative, the state itself is entangled. Finding positive maps, unfortunately, is as difficult as finding the entanglement witnesses.

In the case of two qubits, the criterion involving positive maps becomes very simple. It reduces to the statement that if partial transposition of the bipartite state leads to a negative operator, then the state is entangled. This is known as the Peres–Horodecki criterion.

9
Quantum entanglement in practice

I shall now discuss how entanglement witnesses[1] can be measured in practice. There are many beautiful results in this direction which are also highly relevant techniques in other areas.[2]

9.1 Measurements with a Mach–Zehnder interferometer

The main idea is that the very simple Mach–Zehnder setup that we described earlier can be used to test for and even measure entanglement. I shall first explain the key idea and then discuss its exact application. The main setup is as follows.

Consider a conventional Mach–Zehnder interferometer in which the beam-pair spans a two-dimensional Hilbert space $\mathcal{E}H = \{|\tilde{0}\rangle, |\tilde{1}\rangle\}$. The state vectors $|\tilde{0}\rangle$ and $|\tilde{1}\rangle$ can be taken as (orthogonal) wave packets that move in two given directions defined by the geometry of the interferometer. In this basis, we may represent mirrors, beamsplitters, and relative phase shifts by the unitary operators

$$\tilde{U}_M = \begin{pmatrix} 0 & 1 \\ 1 & 0 \end{pmatrix}, \quad \tilde{U}_B = \frac{1}{\sqrt{2}} \begin{pmatrix} 1 & 1 \\ 1 & -1 \end{pmatrix},$$

$$\tilde{U}(1) = \begin{pmatrix} e^{i\chi} & 0 \\ 0 & 1 \end{pmatrix}, \tag{9.1}$$

respectively. An input pure state $\tilde{\varrho}_{in} = |\tilde{0}\rangle\langle\tilde{0}|$ of the interferometer is transformed into the output state

$$\begin{aligned}\tilde{\varrho}_{out} &= \tilde{U}_B \tilde{U}_M \tilde{U}(1) \tilde{U}_B \tilde{\varrho}_{in} \tilde{U}_B^\dagger \tilde{U}^\dagger(1) \tilde{U}_M^\dagger \tilde{U}_B^\dagger \\ &= \frac{1}{2}\begin{pmatrix} 1+\cos\chi & i\sin\chi \\ -i\sin\chi & 1-\cos\chi \end{pmatrix}\end{aligned} \tag{9.2}$$

this yields the intensity along the direction of $|\tilde{0}\rangle$ as $I \propto 1 + \cos\chi$. Thus the relative phase χ can be observed in the output signal of the interferometer.

Now assume that the particles carry additional internal degrees of freedom, for example spin (this degree of freedom will be transported in parallel). The internal

---
[1] And also various measures of entanglement that will be discussed in detail in the next chapter
...

[2] such as geometric phases, for example, but that is a subject for my next book I suppose!

spin space $\mathcal{E}H_i \cong \mathcal{E}C^N$ is spanned by the vectors $|k\rangle$, $k = 1, 2, \ldots, N$. The density operator can be made to change inside the interferometer in accordance with

$$\varrho_0 \longrightarrow U_i \varrho_0 U_i^\dagger, \tag{9.3}$$

where U_i is a unitary transformation acting only on the internal degrees of freedom (we shall see later on that this transformation need not be unitary, but could be a more general completely positive map). Mirrors and beamsplitters are assumed to leave the internal state unchanged, and so we may replace $\tilde{U}_M$ and $\tilde{U}_B$ by $\mathbf{U}_M = \tilde{U}_M \otimes I_i$ and $\mathbf{U}_B = \tilde{U}_B \otimes I_i$, respectively, I_i being the internal unit operator. Furthermore, we introduce the unitary transformation

$$\mathbf{U} = \begin{pmatrix} 0 & 0 \\ 0 & 1 \end{pmatrix} \otimes U_i + \begin{pmatrix} e^{i\chi} & 0 \\ 0 & 0 \end{pmatrix} \otimes I_i. \tag{9.4}$$

The operators $\mathbf{U}_M$, $\mathbf{U}_B$, and $\mathbf{U}$ act on the full Hilbert space $\mathcal{E}H \otimes \mathcal{E}H_i$. $\mathbf{U}$ corresponds to the application of U_i along the path of $|\tilde{1}\rangle$ and the $U(1)$ phase χ along the path of $|\tilde{0}\rangle$. We shall use $\mathbf{U}$ to generalize the notion of a phase to unitarily evolving mixed states.

Let an incoming state, given by the density matrix $\varrho_{\text{in}} = \tilde{\varrho}_{\text{in}} \otimes \varrho_0 = |\tilde{0}\rangle\langle\tilde{0}| \otimes \varrho_0$, be split coherently by a beamsplitter and recombine at a second beamsplitter after being reflected by two mirrors. Suppose that $\mathbf{U}$ is applied between the first beamsplitter and the mirror pair. The incoming state is transformed into the output state

$$\varrho_{\text{out}} = \mathbf{U}_B \mathbf{U}_M \mathbf{U} \mathbf{U}_B \varrho_{\text{in}} \mathbf{U}_B^\dagger \mathbf{U}^\dagger \mathbf{U}_M^\dagger \mathbf{U}_B^\dagger. \tag{9.5}$$

Inserting eqns 9.1 and 9.4 into eqn 9.5 yields

$$\varrho_{\text{out}} = \frac{1}{4} \left[\begin{pmatrix} 1 & 1 \\ 1 & 1 \end{pmatrix} \otimes U_i \varrho_0 U_i^\dagger + \begin{pmatrix} 1 & -1 \\ -1 & 1 \end{pmatrix} \otimes \varrho_0 \right.$$

$$+ e^{i\chi} \begin{pmatrix} 1 & 1 \\ -1 & -1 \end{pmatrix} \otimes \varrho_0 U_i^\dagger$$

$$\left. + e^{-i\chi} \begin{pmatrix} 1 & -1 \\ 1 & -1 \end{pmatrix} \otimes U_i \varrho_0 \right]. \tag{9.6}$$

The output intensity along $|\tilde{0}\rangle$ is

$$I \propto \text{tr}\left(U_i \varrho_0 U_i^\dagger + \varrho_0 + e^{-i\chi} U_i \varrho_0 + e^{i\chi} \varrho_0 U_i^\dagger \right)$$

$$\propto 1 + |\text{tr}(U_i \varrho_0)| \cos\left[\chi - \arg \text{tr}(U_i \varrho_0) \right], \tag{9.7}$$

where we have used $\text{tr}(\varrho_0 U_i^\dagger) = [\text{tr}(U_i \varrho_0)]^*$.

The important observation from eqn 9.7 is that the interference oscillations produced by the variable $U(1)$ phase χ are shifted by $\phi = \arg \text{tr}(U_i \varrho_0)$ for any internal input state ϱ_0, be it mixed or pure. This phase difference reduces for pure states $\varrho_0 = |\psi_0\rangle\langle\psi_0|$ to the Pancharatnam phase difference between $U_i|\psi_0\rangle$ and $|\psi_0\rangle$. Moreover the visibility of the interference pattern is $\nu = |\text{tr}(U_i \varrho_0)| \geq 0$, which reduces to the expected $\nu = |\langle\psi_0|U_i|\psi_0\rangle|$ for pure states.

The key observation for us is that by suitably choosing the unitary transformation $\mathbf{U}$ we can obtain traces of various powers of a density matrix, $\text{tr}\varrho^k$. From these, any function of a density matrix can be inferred.

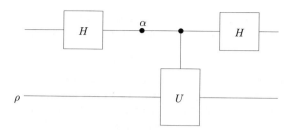

Fig. 9.1 The dot with an alpha next to it represents a phase shift under which $|0\rangle \to |0\rangle$ and $|1\rangle \to e^{i\alpha}|1\rangle$. The gate U connected to a dot represents a control gate, U is executed when the control gate is in the state $|1\rangle$. Thus Control-U can be written as $|0\rangle \otimes \varrho \to \varrho$ and $|1\rangle \otimes \varrho \to |1\rangle \otimes U\varrho$. H represents a Hadamard state. This circuit is related to the Mach–Zehnder interferometer—the two Hadamard gates H correspond to beam splitters, and the other two gates are applied only in the arm of the interferometer which corresponds to $|1\rangle$. $\mathrm{tr}\, U\varrho = v\cdot e^{i\alpha}$, where v represents the reduction in visibility and α represents the shift.

9.2 Interferometric implementation of Peres–Horodecki criterion

Partial transposition is not a physical operation, in that it is a positive map and not a CP-map. Therefore it cannot be implemented directly within the quantum formalism. Looking at a simple 2×2 matrix of a qubit, transposition (in some sense) corresponds to time reversal (the off-diagonal elements are exchanged and this is like $e^{i\omega t} \to e^{-i\omega t}$; we can also think of it as a phase conjugation in general since it corresponds to the transformation $i \to -i$). However, we should remember that an entanglement witness is the average of some Hermitian operator, and this average is a physically measurable quantity. So we must be able to measure the effects of the partial transposition in some indirect way.

The key observation is that we can use an interferometer to measure powers of an operator, $\mathrm{tr}(\varrho^n)$. With a bit more ingenuity we can also measure $\mathrm{tr}((\varrho_{12}^{T_2})^n)$, that is, all the powers of the partially transposed matrix (here, 1 and 2 refer to the two parts of the system, i.e. A and B). These quantities will just be some real (not complex) numbers, so we are not contradicting quantum physics! From this, the spectrum of the partially transposed matrix can be estimated and therefore we can see if any of its eigenvalues are negative (thereby signaling the existence of entanglement). First we shall explain how to measure the trace of the square of a density matrix ϱ. This is sometimes referred to as the purity factor, since it tells us how close to a pure state ϱ is (see Chapters 2 and 3). The maximum of 1 is achieved when ϱ is pure, and the minimum is reached by a maximally mixed state.

9.2.1 Measuring tr ϱ^2?

Bearing in mind that the interferometer naturally measures the quantity $\mathrm{tr}\, U\sigma$, we need to think of the right choice for both the unitary transformation U and the initial state σ in order to produce the trace of the square of some density ϱ. After some thought, we can realize that this can be done by applying a (unitary) swap operator in one arm of the interferometer to an internal state of the form $\varrho \otimes \varrho$. Figure 9.1 shows how this works and the actual circuit is shown in Fig. 9.2.

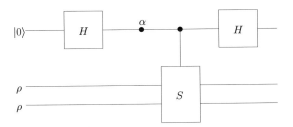

Fig. 9.2 Circuit for computing $\text{tr}(\varrho^2)$. The components are defined as in Fig. 9.1, and S is a swap gate.

The state swap operation is defined as

$$S|\phi\rangle|\psi\rangle = |\psi\rangle|\phi\rangle \tag{9.8}$$

for any two states $|\phi\rangle$ and $|\psi\rangle$. We can now easily confirm that

$$\text{tr}\, S\varrho \otimes \varrho = \text{tr}\, \varrho^2 \tag{9.9}$$

since when we swap all the elements in the bra part of the density matrices, only the diagonal elements survive after the trace operation. I leave this to the reader to confirm in a more rigorous fashion.

9.2.2 Generalization to $\text{tr}\, \varrho^k$

The operation above can easily be generalized to any power of the density matrix, so that we can always infer $\text{tr}\varrho^n$ from just the interferometric measurements described thus far. The key observation is that we need a generalization of the two-state swap operation to n states. This generalized operation is easily seen to be

$$S^{(k)}|\phi_1\rangle|\phi_2\rangle\ldots|\phi_k\rangle = |\phi_k\rangle|\phi_1\rangle\ldots|\phi_{k-1}\rangle . \tag{9.10}$$

Thus

$$\text{tr}\, S^{(k)} \varrho_1 \otimes \ldots \varrho_k = \text{tr}(\varrho_1\varrho_2\ldots\varrho_k) \Rightarrow \text{tr}\, \varrho^k .$$

The main reason why we would like to have all the powers of a density matrix is that we can infer all its eigenvalues from this information. In fact, not all powers are needed, just as many as the dimensionality of the state.

This is done in the following way. Knowing the powers of ϱ means that we know the sum of the powers of the eigenvalues of the density matrix:

$$e_1^k + e_2^k + \ldots e_n^k = \text{tr}\varrho^k . \tag{9.11}$$

From this set of n linearly independent equations we can infer all n eigenvalues (for $k = 1$ this is, of course, just a trivial statement of the normalization of the density matrix). Note that the trace of any function of the density matrix can also be estimated in this way, since

$$\text{tr}f(\varrho) = \sum_i f(e_i) \tag{9.12}$$

and $f(e_i)$ can easily be computed once the e_i's are all known. This also means that the entropy can be computed in the same simple way.

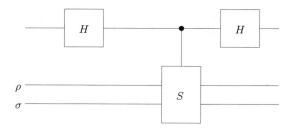

Fig. 9.3 Circuit for computing the fidelity between ϱ and σ. The components are defined as in Fig. 9.1.

9.2.3 Measuring tr $(\varrho^{T_2})^k$

The simple strategy to check for entanglement of two qubits now is just to try to estimate the eigenvalues of $\varrho_{12}^{T_2}$. For this we need to be able to compute $\text{tr}(\varrho^{T_2})^k$ for all k's up to the dimensionality of the state.

There is a simple method of measuring the eigenvalues of the partially transposed state and therefore of checking the Peres–Horodecki criterion. It uses the same interferometric setup and the method rests on the following observation:[3]

$$\text{tr}(\varrho^{T_2})^2 = \text{tr}\{S_{13}S_{24}\varrho_{12} \otimes \varrho_{34}\} . \tag{9.13}$$

Therefore we now need two swap operations to compute the trace of the square of the partially transposed state. Higher powers of the partial transpose can be obtained by a simple generalization, and therefore all the eigenvalues of the partially transposed matrix can be estimated. If one of them is negative, then we can conclude that the original density matrix is entangled.

Other interesting quantities can also be computed using the Mach–Zehnder interferometer. We show how the fidelity arises within this context.

9.3 Measuring the fidelity between ϱ and σ

So far, we have analyzed how to compute various functions of a state. But the interferometric setup is also suitable for measuring and comparing different states. For example, as we have seen, the fidelity between two mixed states can be defined as follows:

$$F = \text{tr}(\varrho\sigma) = \text{tr}(\sigma\varrho) . \tag{9.14}$$

To measure this, we again need to be able to implement the state swap operation discussed above, $V|i\rangle|j\rangle = |j\rangle|i\rangle$, and this leads to the relation $\text{tr}\, V(\sigma \otimes \varrho) = \text{tr}\,\sigma\varrho$. Therefore, the trace of the product of two density matrices can easily be measured. Thus we can also measure the mutual information, the relative entropy and all the other measures between states introduced in this book.

We have now finished with the Mach–Zehnder setup and its role in the estimation and witnessing of entanglement. The next question we address is the following: can an initially nonmaximally entangled state (we consider pure states and then mixed states)

---

[3]This very beautiful result is due to Hilary Carteret.

be converted into a maximally entangled state (a singlet, say)? From this question, we shall derive various different ways of quantifying entanglement and then discuss some intriguing connections with thermodynamics.

9.4 Summary

The important message of this chapter is that a simple apparatus that measures quantum superpositions, such as a Mach–Zehnder interferometer, can also be used for much more complicated measurements. The key quantity that is measured by an interferometer is

$$\operatorname{tr}(U\varrho) , \tag{9.15}$$

where U is a unitary transformation implemented in one arm of the interferometer on the state ϱ. We have all possible freedom in choosing both U and ϱ.

This expression and the above setup can be used to measure the trace of any function of the density matrix and of various functions of two (or more) density matrices. Consequently, entanglement witnesses can also be measured (or implemented) in the same way. Physically, this is because once we can utilize single-particle superpositions, we can build up any other more complicated superposition from these. Paraphrasing Feynman, there is only one mystery in quantum mechanics (the superposition principle), and all other mysteries follow from it.

10
Measures of entanglement

We have seen in the previous chapters how to tell whether two states are entangled or not. We would now like to be able to tell how entangled two states are. One obvious way is to consider the distillation procedure—we can quantify the entanglement of any state by how many fully entangled qubits can be distilled from an entangled state. However, it is not easy to see how to distil singlets out of n copies of a general mixed state.[1] So, let us first see what can be done with pure states.

10.1 Distillation of multiple copies of a pure state

The general scenario here is that Alice and Bob initially share n entangled qubit pairs, where each pair is prepared in the state

$$|\psi^{AB}\rangle = \sqrt{p}|0^A 0^B\rangle + \sqrt{1-p}|1^A 1^B\rangle , \quad (10.1)$$

where p is real, though this is no fundamental restriction in practice. We can write the joint system of n pairs as

$$\begin{aligned}
|\psi^{AB}\rangle^{\otimes n} &= (\sqrt{p}|0^A 0^B\rangle + \sqrt{1-p}|1^A 1^B\rangle) \otimes \ldots \otimes (\sqrt{p}|0^A 0^B\rangle + \sqrt{1-p}|1^A 1^B\rangle) \\
&= p^{n/2}|0^A 0^B 0^A 0^B \ldots 0^A 0^B\rangle \\
&+ p^{(n-1)/2}(1-p)^{n/2}(|0^A 0^B 0^A 0^B \ldots 1^A 1^B\rangle + \ldots \\
&+ |1^A 1^B 0^A 0^B \ldots 0^A 0^B\rangle) + \ldots \\
&+ (1-p)^{n/2}|1^A 1^B 1^A 1^B \ldots 1^A 1^B\rangle \\
&= \sum_{x_1 \ldots x_n \in \{0,1\}^n} \sqrt{p}^i \sqrt{1-p}^{n-i} |x_1 \ldots x_n\rangle^A \otimes |x_1 \ldots x_n\rangle^B .
\end{aligned} \quad (10.3)$$

Let $\varrho = p|0\rangle\langle 0| + (1-p)|1\rangle\langle 1|$. Alice and Bob each have the mixed state

$$\varrho^A = \varrho^B = \sum_i p^i (1-p)^{n-i} |i\rangle\langle i| = \varrho^{\otimes n} . \quad (10.4)$$

In studying data compression in Chapters 1 and 5, we saw that, as n grows large:

---

[1] I'll be honest about it: we actually do not have a clue how to do this.

- A projection can be performed onto the typical subspace of $\varrho^{\otimes n}$ with arbitrarily high probability of success. Alice and Bob each can perform local projections onto the typical subspace and transform the joint system to

$$\sum_{x_1\ldots x_n \in \mathrm{Typ}(\varrho)} \sqrt{p}^i \sqrt{1-p}^{n-i} |x_1 \ldots x_n^A\rangle \otimes |x_1 \ldots x_n^B\rangle \quad (10.5)$$

with arbitrary probability of success.
- For any $\epsilon > 0$ and $\delta > 0$, the probability of each string $|x_1 \ldots x_n\rangle$ in the typical subspace satisfies

$$2^{-n(H(X)-\delta)} < p(x_1, \ldots, x_n) < 2^{-n(H(X)+\delta)}, \quad (10.6)$$

and so the projected state is arbitrarily close to

$$\sum_{x_1\ldots x_n \in \mathrm{Typ}(\varrho)} 2^{-nH(X)} |x_1 \ldots x_n\rangle. \quad (10.7)$$

- The number of strings in the typical subspace is arbitrarily close to $2^{nH(X)}$. Alice and Bob can unitarily transform the projected state (with arbitrary fidelity) to

$$|\phi\rangle = \sum_{i=1}^{nH(X)} \sqrt{2^{-nH(X)}} |i^A\rangle |i^B\rangle. \quad (10.8)$$

In the above, $|\phi\rangle$ is the distilled state of $|\psi^{AB}\rangle$ obtained using only LOCC. Alice and Bob distill $|\psi^{AB}\rangle^{\otimes n}$ to a maximally entangled state of $nS(\varrho)$ qubits. And this is the whole point of distillation: we would like to convert some number of nonmaximally entangled states into a (necessarily smaller) number of maximally entangled states. The final state is effectively given by $nH(X)$ maximally entangled two-qubit states.

The same protocol can be applied to initially mixed states shared between Alice and Bob, however, the rate of conversion to maximum entanglement cannot then be known in general (as we shall discuss in more detail later).

Note also that the converse process of distillation is possible. Namely, we can start from a set of maximally entangled states and locally convert them into a (greater) number of nonmaximally entangled states. This is known as formation of entanglement. The distillation and formation procedures provide us with two ways of quantifying entanglement:

- The entanglement of distillation E_D is defined as $\lim_{n\to\infty} m(n)/n$, which represents the number m of maximally entangled states that can be derived from an initial number n of nonmaximally entangled states using only LOCC.
- The entanglement of formation E_F is defined as the minimum number of maximally entangled states that one must invest in order to obtain a certain number of copies of a given state by LOCC (again defined in the asymptotic sense).

We shall discuss these two measures in greater detail later, but first we would like to explore their analogy with Carnot cycles in thermodynamics.

10.2 Analogy with the Carnot Cycle

We have seen that entanglement cannot increase under LOCC. Physically this makes sense since entanglement is a nonlocal property (according to Bell's notion of nonlocality) and should not be increased by just local manipulations. This rule is reminiscent of the Second Law of Thermodynamics, which says that entropy cannot decrease under adiabatic processes. We can make an analogy between entanglement and entropy on the one hand and LOCC and adiabatic processes on the other hand. Both laws serve to prohibit certain types of processes which are impossible in nature—decrease in entropy with adiabatic processes and causing an increase in entanglement with LOCC. This analogy is useful in terms of seeing how different physical properties relate to one another, and we can use it derive the distillation and formation entanglements for pure states from it.[2]

We begin by restating the Second Law of Thermodynamics in a form due to Clausius:

There exists no thermodynamic process the sole effect of which is to extract a quantity of heat from the colder of two reservoirs and deliver it to the hotter of the two reservoirs.

Suppose now that we have a thermodynamic system. We want to invest some heat in it so that, at the end, our system does as much work as possible with this heat input and returns to its initial state (so that we can repeat the cycle and use the system again!). The efficiency of the system is defined as the rate of the output work W_{out} to the invested heat Q_{in}:

$$\eta = \frac{W_{\text{out}}}{Q_{\text{in}}} \tag{10.9}$$

It is well-known that the efficiency is maximized if we have a reversible process—any friction leading to irreversibility will waste useful work.

An example of a process in which the efficiency is known is the Carnot cycle. We can show that the most efficient way to run a Carnot cycle is reversibly. The argument is a *reductio ad absurdum* which can be found in most undergraduate texts on thermodynamics. I shall reiterate it here for the reader's convenience. A Carnot engine is made up of two reservoirs—a hotter reservoir and a colder reservoir. The aim of a Carnot engine C is to take some heat input from a hotter reservoir, do as much work with it as possible and deliver some heat to the colder reservoir. If the Carnot engine is reversible and if we run the Carnot engine forwards and then backwards, the net result is no change in heat. Suppose there exists another reversible machine E which we suppose can perform more work than the Carnot engine. If we run E and then C in reverse, the net result is to extract heat from the colder reservoir and deliver it to the hotter reservoir. Extracting heat from the colder system and delivering it to the hotter one contradicts the Second Law.

We shall use the Carnot cycle argument to guide us as we try to show that the pure-state distillation described earlier in the chapter is optimal.[3] This distillation scheme is "reversible" since the original state $|\psi^{AB}\rangle^{\otimes n}$ can be obtained with arbitrary fidelity from the final state $|\Phi^+\rangle^{\otimes m}$, where $m = nS(\varrho)$ (this will be seen in more detail

---

[2] We shall have much more to say later in this chapter about the formal analogy between thermodynamics and entanglement.

[3] Optimality is already suggested by the fact that its efficiency is given by the entropy.

later when we discuss the formation of entanglement). The efficiency of the scheme is the number of fully entangled pairs $|\Phi^+\rangle$ which are obtained. Suppose that there was a more efficient reversible process E that transformed the state $|\psi^{AB}\rangle^{\otimes n}$ into $M > m$ fully entangled pairs. Then if we run E forwards followed by reversing the original process we obtain $|\psi^{AB}\rangle^{\otimes M/S(\varrho)}$, where $M/S(\varrho) > n$ is the number of copies of the original state. By performing two LOCC operations, we obtain $M/S(\varrho) - n > 0$ copies of the partially entangled state $|\psi^{AB}\rangle$, which violates the law that entanglement cannot be increased under LOCC.[4]

We have thus shown that the fact that LOCC cannot increase entanglement can be taken as a very fundamental principle in quantum information theory. When this is applied to mixed states, however, the results are not as clear as for pure states. In fact, there is a fundamental irreversibility involved in understanding local manipulation of mixed states. This is the subject of the remainder of the chapter.

10.3 Properties of entanglement measures

A good place to start with quantifying entanglement is to consider what properties we would like our entanglement measure to possess. However, unlike when information is quantified, we shall not be able to find a unique measure in general for mixed states. Different measures will be useful in different scenarios.

The first property we need from an entanglement measure is that a disentangled state does not have any quantum correlations. This gives rise to our first condition:

E1. For any separable state σ, the measure of entanglement should be zero, that is,

$$E(\sigma) = 0 . \tag{10.10}$$

Note that we do not necessarily require the converse, that is, that if $E(\sigma) = 0$, then σ is separable. The reason for this will become clear below and has to do with which type of entanglement we would like to investigate.

The next condition, E2, concerns the behavior of the entanglement under simple local unitary transformations. A local unitary transformation represents a change of the basis in which we consider the given entangled state. But a change of basis should not change the amount of entanglement that is accessible to us, because at any time we could just reverse the basis change (since unitary transformations are fully reversible).

E2. For any state σ and any local unitary transformation, that is, a unitary transformation of the form $U_A \otimes U_B$, the measure of entanglement remains unchanged. Therefore

$$E(\sigma) = E(U_A \otimes U_B \sigma U_A^\dagger \otimes U_B^\dagger) . \tag{10.11}$$

The third condition is the one that really restricts the class of possible entanglement measures. Unfortunately, it is usually also the property that is the most difficult to prove for potential measures of entanglement. We have already argued that no good measure of the correlation between two subsystems should increase under local operations on the subsystems performed separately. However, quantum entanglement is

---
[4]This argument is basically due to Popescu and Rohrlich (1997).

112 *Measures of entanglement*

even more restrictive in that the total amount of entanglement cannot increase *even with the aid of classical communication*. Classical correlation, on the other hand, can be increased by LOCC.

Example. Suppose that Alice and Bob share n uncorrelated pairs of qubits, for example all in the state $|0\rangle$. Alice's computer then interacts with each of her qubits such that it randomly flips each qubit with probability $1/2$. However, whenever a qubit is flipped, Alice's computer (classically) calls Bob's computer and informs it to do likewise. After this action on all the qubits, Alice and Bob end up sharing n (maximally) correlated qubits in the state $|00\rangle\langle 00| + |11\rangle\langle 11|$, that is, whenever Alice's qubit is zero so is Bob's, and whenever Alice's qubit is one so is Bob's. The state of each pair is mixed because Alice and Bob do not know whether their computers have flipped their respective qubits or not.

We can always calculate the total amount of entanglement by summing the measure of the entanglement of all systems after we have applied our local operations and classical communications.

Our third condition is the following:

E3. Local operations, classical communication, and subselection cannot increase the expected entanglement, that is, if we start with an ensemble in a state σ and end up, with probability p_i in subensembles in a state σ_i, then we shall have

$$E(\sigma) \geq \sum_i p_i E(\sigma_i) \,, \qquad (10.12)$$

where $\sigma_i = A_i \otimes B_i \sigma A_i^\dagger \otimes B_i^\dagger / p_i$ and $p_i = \mathrm{tr}(A_i \otimes B_i \sigma A_i^\dagger \otimes B_i^\dagger)$. The form $A \otimes B$ shows that Alice and Bob perform their operations locally (i.e. Alice cannot affect Bob's system and vice versa). However, Alice's and Bob's operations can be classically correlated which is manifested in the fact that they have the same index. It should be pointed out that although all LOCC can be cast in the above product form, the converse is not true: not all operations of product form can be executed locally. This means that the above condition is more restrictive than necessary, but this does not have any significant consequences as far as I am aware. An example of an E3 operation is local addition of particles on Alice's and Bob's sides. Note also that E2 operations are a subset (a special case) of E3 operations.[5]

Before I introduce various entanglement measures I would like to discuss the following question: "What do we mean by saying that a state σ can be converted into another state ϱ by LOCC?". Strictly speaking, we mean that there exists an LOCC operation that, given a sufficiently large number n of copies of σ, will convert them arbitrarily close to m copies of the state ϱ, that is,

$$(\forall \epsilon > 0)(\forall m \in N)(\exists n \in N; \exists \Phi \in \mathrm{LOCC}) \,,$$
$$||\Phi(\sigma^{\otimes n}) - \varrho^{\otimes m}|| < \epsilon \,, \qquad (10.13)$$

where $||\sigma - \varrho||$ is some measure of distance (metric) on the set of density matrices (say 1 minus the fidelity). Now, if σ is more entangled than ϱ, we expect that there will be

---

[5]Readers interested in further analysis of these conditions are advised to read Vidal (2000) and Horodecki et al. (2000).

an LOCC operation such that $m > n$; otherwise, we expect that we can have $n \leq m$. Measuring entanglement now reduces to finding an appropriate function on the set of states to order them according to their local convertibility. This is usually achieved by letting either σ or ϱ be a maximally entangled state.

We shall now try to find functions of density matrices that satisfy the above properties. First we show that for pure states, we can easily argue for a unique measure.

10.4 Entanglement of pure states

Given a pure state $|\psi^{AB}\rangle$, we can take the Schmidt decomposition

$$|\psi^{AB}\rangle = \sum_i \alpha_i |i^A i^B\rangle \qquad (10.14)$$

and use a function of the α_i's to quantify entanglement. We require the following conditions for pure-state entanglement measures:

1. E is invariant under local unitary operations, and so E is a function of the α_i's only.
2. E is continuous—this is a natural assumption for physical quantities.
3. E is additive—$E(|\psi^{AB}\rangle \otimes |\phi^{AB}\rangle) = E(|\psi^{AB}\rangle) + E(|\psi^{AB}\rangle)$.

These are exactly the conditions which we used to define the Shannon entropy in Chapter 1—so our entanglement measure of pure states is also an entropy. The unique measure of entanglement (up to an affine transformation) for the pure state $|\psi^{AB}\rangle$ is given by

$$S(\operatorname{tr}_A(|\psi^{AB}\rangle)) = -\sum_i |\alpha_i|^2 \log(|\alpha_i|^2), \qquad (10.15)$$

and this is just the Shannon entropy of the moduli squared of the Schmidt coefficients. In other words, how entangled two systems in a pure state are is given by the von Neumann entropy of (either of) the reduced density matrices. Note that we shall not require conditions 2 and 3 above to hold for mixed states. The reason for this will become transparent shortly.

10.5 Entanglement of mixed states

The Schmidt decomposition, unfortunately, does not apply to mixed states, so the above measure of entanglement will not suffice for mixed states. Consider two mixed states

$$\varrho_E = \frac{1}{4}(|00\rangle\langle 00| + |00\rangle\langle 11| + |11\rangle\langle 00| + |11\rangle\langle 11|) = |\Phi^+\rangle\langle\Phi^+|, \qquad (10.16)$$

$$\varrho_D = \frac{1}{2}(|00\rangle\langle 00| + |11\rangle\langle 11|) = |\Phi^+\rangle\langle\Phi^+| + |\Phi^-\rangle\langle\Phi^-|. \qquad (10.17)$$

Here ϱ_E is a fully entangled state of two qubits; however, ϱ_D, which is an equal mixture of two fully entangled states, is completely disentangled. Both have the same entropy for the reduced density matrices, and so they cannot be distinguished in this way. It turns out, rather fortunately, that there are several natural ways of measuring the entanglement of mixed states according to the conditions above.

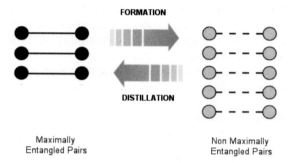

Fig. 10.1 This figure illustrates the formation of entangled states: a certain number of maximally entangled EPR pairs is manipulated by local operations and classical communication and converted into pairs in some state ϱ. The asymptotic conversion rate is known as the entanglement of formation. The converse of formation is the distillation of entanglement. The asymptotic rate of conversion of pairs in the state ϱ into maximally entangled states is known as the entanglement of distillation. The two measures of entanglement are in general different, the entanglement of distillation being greater than or equal to that of formation. The surprising irreversibility of entanglement conversion is a consequence of loss of classical information about the decomposition of ϱ.

I shall now introduce three different measures of entanglement, all of which obey E1–E3. First I discuss the entanglement of formation (see Fig. 10.1). The entanglement of formation of a state ϱ is defined by **Entanglement of formation.**

$$E_{\mathrm{F}}(\varrho) := \min \sum_i p_i S(\varrho_i^A) \,, \tag{10.18}$$

where $S(\varrho^A) = -\operatorname{tr} \varrho^A \log \varrho^A$ is the von Neumann entropy, and the minimum is taken over all the possible realizations of the state $\varrho^{AB} = \sum_j p_j |\psi_j\rangle\langle\psi_j|$, where $\varrho_i^A = \operatorname{tr}_B(|\psi_i\rangle\langle\psi_i|)$. This measure satisfies E1, E2 and E3. The basis of the formation is that Alice and Bob would like to create an ensemble of n copies of the nonmaximally entangled state ϱ_{AB}, using only local operations, classical communication, and a number, m, of maximally entangled pairs (see Fig. 10.1). The entanglement of formation is the asymptotic conversion rate m/n in the limit of infinitely many copies. The form of this measure given in eqn 10.18 will be more transparent after the next subsection, on the entanglement of distillation. Furthermore, I shall analyze the relationship between the entanglement of formation and other measures in more detail later. It is worth mentioning that a closed form for this measure exists for two qubits.
Entanglement of distillation. This measure defines the amount of entanglement of a state σ as the asymptotic proportion of singlets that can be distilled using a purification procedure. This is the opposite process to that leading to the entanglement of formation (Fig. 10.1), although the value obtained is generally smaller. This implies that the formation of states is in some sense irreversible. The reason for this irreversibility will be explained in due course. This measure fails to satisfy the converse of E1: namely, for all disentangled states, the entanglement of distillation is zero,

but the converse is not true. There exist states which are entangled but for which no entanglement can be distilled from them, and, for this reason, they are called *bound entangled* states. This is the reason why the condition E1 was not stated as both a necessary and a sufficient condition.

I shall now introduce the final measure of entanglement, which was first proposed by myself and my colleagues. This measure is intimately related to the entanglement of distillation by providing an upper bound for it, but it can also be related to the entanglement of formation and other measures of entanglement. If $\mathcal{E}D$ is the set of all disentangled states, the measure of entanglement for a state σ is defined as follows:

Relative entropy of entanglement.

$$E(\sigma) := \min_{\varrho \in \mathcal{E}D} S(\sigma||\varrho) \qquad (10.19)$$

where $S(\sigma||\varrho)$ is the quantum relative entropy. This measure, which I shall call the relative entropy of entanglement, tells us that the amount of entanglement in σ is its distance from the disentangled set of states (see Fig. 10.2). In the statistical terms introduced in Chapter 5, the more entangled a state is, the more distinguishable it is from a disentangled state.[6] To understand better all three measures of entanglement we need to introduce another quantum protocol that relies fundamentally on entanglement.

Another condition that might be considered intuitive for a measure of entanglement is *convexity*. Namely, we might require that

$$E\left(\sum_i p_i \sigma^i\right) \leq \sum_i p_i E(\sigma^i) .$$

This states that mixing cannot increase entanglement. For example, an equal mixture of two maximally entangled states $|00\rangle + |11\rangle$ and $|00\rangle - |11\rangle$ is a separable state and consequently contains no entanglement. I did not include convexity as a separate requirement for an entanglement measure, as it is not completely independent of E3. This is because E3 and strong additivity ($E(\varrho \otimes \sigma) = E(\varrho) + E(\sigma)$) imply convexity:

$$n \sum_i p_i E(\varrho_i) = E(\varrho_1^{\otimes p_1 n} \varrho_2^{\otimes p_2 n} \ldots \varrho_N^{\otimes p_N n})$$

$$\geq E\left(\left(\sum_i p_i \varrho_i\right)^{\otimes n}\right) = nE\left(\sum_i p_i \varrho_i\right),$$

where the equalities follow from the assumption of strong additivity and the inequality is a consequence of E3. The symbol $\varrho^{\otimes m}$ means (as always) that we have m copies of the state ϱ. Nevertheless, it is interesting to point out that any convex measure that satisfies continuity and weak additivity has to be bounded from below by the entanglement of distillation and from above by the entanglement of formation. We shall

---

[6]This quantity is not additive which is why we do not require additivity, of entanglement measures for mixed states, unlike the case for pure states.

116 Measures of entanglement

see that "most" entanglement measures can in fact be generated using the quantum relative entropy.

It is interesting to note that the relative entropy of entanglement does in fact satisfy both convexity and continuity, although not additivity. Furthermore, we can easily show that it is an upper bound on the entanglement of distillation. For any pure state $|\psi\rangle$, $\min_{\omega \in \mathcal{E}D} S(\psi^{\otimes n}||\omega) = \min_{\omega \in \mathcal{E}D} -\langle\psi^{\otimes n}|\log\omega|\psi^{\otimes n}\rangle$. However, the logarithmic function is concave so that

$$\min_{\omega \in \mathcal{E}D} -\langle\psi^{\otimes n}|\log\omega|\psi^{\otimes n}\rangle \geq \min_{\omega \in \mathcal{E}D} -\log\langle\psi^{\otimes n}|\omega|\psi^{\otimes n}\rangle \tag{10.20}$$

However, since ω is a disentangled state, its fidelity with the maximally entangled state cannot be larger than the inverse of half the dimension of that state,[7] so that $\langle\psi^{\otimes n}|\omega|\psi^{\otimes n}\rangle \leq 1/2^n$. Thus,

$$\min_{\omega \in \mathcal{E}D} S(\psi^{\otimes n}||\omega) \geq -\log\left(\frac{1}{2^n}\right) = n\ . \tag{10.21}$$

---

[7]For two qubits, this is easily seen from the fact that any state with a fidelity greater than a half can be distilled by LOCC to Werner states with a fidelity greater than a half. These states, as we have seen, are inseparable.

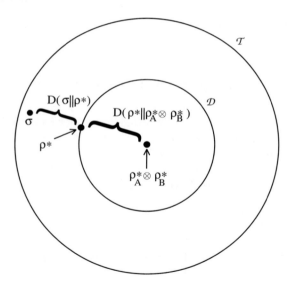

Fig. 10.2 The set of all density matrices, $\mathcal{E}T$, is represented by the outer circle. A subset of it, a set of disentangled states $\mathcal{E}D$, is represented by the inner circle. A state σ belongs to the entangled states, and ϱ^* is the disentangled state that minimizes the distance $D(\sigma||\varrho)$, thus representing the amount of quantum correlation in σ. The state $\varrho_A^* \otimes \varrho_B^*$ is obtained by tracing ϱ^* over A and B. $D(\varrho^*||\varrho_A^* \otimes \varrho_B^*)$ represents the classical part of the correlation in the state σ.

But we know that this minimum is achievable by the state $\omega = \varrho^{\otimes n}$, where ϱ is obtained from $|\psi\rangle$ by removing the off-diagonal elements in the Schmidt basis. Consequently, if we are starting with n copies of the state σ, and obtaining m copies of $|\psi\rangle$ by LOCC, then

$$D = \frac{m}{n} = \frac{1}{n} \min_{\omega \in \mathcal{E}D} S(\psi^{\otimes m}||\omega) \leq \frac{1}{n} \min_{\omega \in \mathcal{E}D} S(\sigma^{\otimes n}||\omega) \quad (10.22)$$

where the equality follows from eqn 10.21 and the inequality from the fact that the relative entropy is nonincreasing under LOCC (strictly speaking, $D = \lim_{n\to\infty} m/n$ and, of course, m is a function of n, i.e. $m = m(n)$). Thus, the distillable entanglement is bounded from above by the relative entropy of entanglement.

A similar argument can be given to show that the relative entropy of entanglement E_{RE} is bounded from the above by the entanglement of formation. It goes as follows:

$$S\left(\sum_i p_i \sigma_i || \omega\right) \leq \sum_i p_i S(\sigma_i || \omega) \leq E_{\text{F}} \quad (10.23)$$

where $S\left(\sum_i p_i \sigma_i || \omega\right) \geq E_{\text{RE}}$, by definition.

Since most of the measures of entanglement can be derived from the relative entropy, they will possess a similar property. The relative entropy is a good measure of entanglement not only because it is a good upper bound on the distillable entanglement and because of its universal appeal, but also because other measures of entanglement can in fact be formally derived from it. We show how this works in the following section.

10.6 Measures of entanglement derived from relative entropy

Suppose that Alice and Bob share a state described by a density matrix ϱ^{AB}. The state ϱ^{AB} has an infinite number of different decompositions $\varepsilon = \{|\psi_i^{AB}\rangle\langle\psi_i^{AB}|, p_i\}$ into pure states $|\psi_i^{AB}\rangle$, with probabilities p_i. We denote the mixed state ϱ^{AB}, written in decomposition ε, by

$$\varrho_\varepsilon^{AB} = \sum_i p_i |\psi_i^{AB}\rangle\langle\psi_i^{AB}|. \quad (10.24)$$

As we have seen, measures of entanglement are associated with the formation and distillation of pure and mixed entangled states. The known relationships between the various measures of entanglement for mixed states are $E_D(\varrho_{AB}) \leq E_{\text{RE}}(\varrho_{AB}) \leq E_F(\varrho_{AB})$. Equality holds for pure states, where all the measures reduce to the von Neumann entropy $S(\varrho_A) = S(\varrho_B)$.

The formation of an ensemble of n nonmaximally entangled *pure* states $\varrho_{AB} = |\psi_{AB}\rangle\langle\psi_{AB}|$ is achieved by the following protocol. Alice first prepares the states she would like to share with Bob locally. She then uses data compression to compress subsystem B into $nS(\varrho_B)$ states. Subsystem B is then teleported to Bob using $nS(\varrho_B)$ maximally entangled pairs. Bob decompresses the states he receives and so ends up sharing n copies of ϱ_{AB} with Alice. The entanglement of formation is therefore $E_F(\varrho_{AB}) = S(\varrho_B)$. For pure states, this process requires no classical communication in the asymptotic limit. The reverse process of distillation is accomplished using the

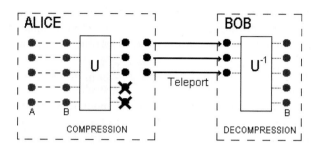

Fig. 10.3 Formation of a state by LOCC. First, Alice creates a joint state of subsystems A and B locally. Then, she performs quantum data compression on the subsystem B and teleports the compressed state to Bob. Finally, Bob decompresses the received state. Alice and Bob end up sharing the joint state of A and B initially prepared by Alice.

Schmidt projection method, which allows $nS(\varrho_B)$ maximally entangled pairs to be distilled in the limit as n becomes very large. No classical communication between the separated parties is required. Therefore pure states are fully interconvertible in the asymptotic limit.

The situation for mixed states is more complex. When any mixed state, as denoted by eqn 10.24, is created, it may be imagined to be part of an extended system whose state is pure. The pure states $|\psi^i_{AB}\rangle$ in the mixture may be regarded as being correlated with orthogonal states $|m_i\rangle$ of a memory M. The extended system is in the pure state $|\psi_{MAB}\rangle = \sum_i \sqrt{p_i}|m_i\rangle|\psi^i_{AB}\rangle$. If we have no access to the memory system, we trace over it to obtain the mixed state in eqn 10.24. In fact, the lack of access to the memory is of a completely general nature. It may be due to interaction with another inaccessible system, or it may be due to an intrinsic loss of information. The results I shall present are universally valid and do not depend on the nature of the information loss. We shall see that the amount of entanglement involved in the various manipulations of the entanglement of mixed states depends on the accessibility of the information in the memory at various stages. Note that a unitary operation on $|\psi_{MAB}\rangle$ will convert it into another pure state $|\phi_{MAB}\rangle$ with the same entanglement, and tracing over the memory yields a different decomposition of the mixed state. Reduction of the pure state to the mixed state may be regarded as due to a projection-valued measurement on the memory with operators $\{E_i = |m_i\rangle\langle m_i|\}$.

Consider first the protocol of formation by means of which Alice and Bob come to share an ensemble of n mixed states ϱ_{AB}. Alice first creates the mixed states locally by preparing a collection of n states in a particular decomposition, $\varepsilon = \{|\psi^i_{AB}\rangle\langle\psi^i_{AB}|, p_i\}$ by making np_i copies of each pure state $|\psi^i_{AB}\rangle$. At the same time we may imagine a memory system entangled with the pure states to be generated, which keeps track of the identity of each member of the ensemble. I consider first the case where the state of subsystems A and B together with the memory is pure. Later, I shall consider the situation in which Alice's memory is decohered. There are three ways for her to share the states with Bob. First of all, she may simply compress subsystem B to $nS(\varrho_B)$ states, and teleport these to Bob using $nS(\varrho_B)$ maximally entangled pairs. The choice of which subsystem to teleport is made so as to minimize the amount of entanglement required, so that $S(\varrho_B) \leq S(\varrho_A)$. The teleportation in this case would require no classical communication in the asymptotic limit, just as for pure states.[8] The state of the whole system which is created by this process is an ensemble of pure states $|\psi_{MAB}\rangle$, where the subsystems M and A are on Alice's side and subsystem B is on Bob's side. In terms of entanglement resources, however, this process is not the most efficient way for Alice to send the states to Bob. She may do this more efficiently by using the memory system of $|\psi_{MAB}\rangle$ to identify blocks of np_i members in each pure state $|\psi_{AB}^i\rangle$, and applying compression to each block to give $np_i S(\varrho_B^i)$ states. The total number of maximally entangled pairs required to teleport the states to Bob is $n \sum_i p_i S(\varrho_B^i)$, which is clearly less than $nS(\varrho_B)$, by concavity of the entropy. The amount of entanglement required clearly depends on the decomposition of the mixed state ϱ_{AB}. In order to decompress these states, Bob must also be able to identify which members of the ensemble are in which state. Therefore Alice must also send him the memory system. She now has two options. She may either teleport the memory to Bob, which would use more entanglement resources, or she may communicate the information in the memory classically, with no further use of entanglement. When Alice uses the minimum-entanglement decomposition $\varepsilon = \{|\psi_{AB}^i\rangle\langle\psi_{AB}^i|, p_i\}$, the latter process makes the most efficient use of entanglement, consuming only the entanglement of formation of the mixed state $E_F(\varrho_{AB}) = \sum_i p_i S(\varrho_B^i)$. We may think of the classical communication between Alice and Bob in either of two equivalent ways. Alice may either measure the memory locally to decohere it, and then send the result to Bob classically, or she may send the memory through a completely decohering quantum channel. Since Alice and Bob have no access to the channel, the state of the whole system which is created by this process is the mixed state

$$\varrho_{ABM}^\varepsilon = \sum_i p_i |\psi_{AB}^i\rangle\langle\psi_{AB}^i| \otimes |m_i\rangle\langle m_i|, \quad (10.25)$$

where Bob is classically correlated with the subsystem AB. Bob is then able to decompress his states using the memory to identify members of the ensemble.

Once the collection of n pairs has been shared between Alice and Bob, it is converted into an ensemble of n mixed states ϱ_{AB} by destroying the access to the memory which contains the information about the state of any particular member of the ensemble. *It is the loss of this information which is responsible for the fact that the entanglement of distillation is lower than the entanglement of formation, since this information is not available to the parties carrying out the distillation.* If Alice and Bob, who do have access to the memory, were to carry out the distillation, they could obtain as much entanglement from the ensemble as was required to form it. In the case where Alice and Bob share an ensemble of the pure state $|\psi_{MAB}\rangle$, they would simply apply the Schmidt projection method. The relative entropy of entanglement gives an upper bound on the distillable entanglement, $E_{\rm RE}(|\psi_{(MA):B}\rangle\langle\psi_{(MA):B}|) = S(\varrho_B)$, which is the same as the amount of entanglement required to create the ensemble of pure states, as described above. Here MA and B are spatially separated subsystems on which joint operations may not be performed. In my notation, I use a colon to separate the local subsystems.

---

[8] Proven by Lo and Popescu (1999). The details of their proof are not relevant to our argument here.

On the other hand, if Alice uses the least possible entanglement to produce an ensemble of the mixed state ϱ_{AB}, together with classical communication, the state of the whole system is an ensemble of the mixed state $\varrho_{ABM}^{\varepsilon}$, and the process is reversible. Because of the classical correlation with the states $|\psi_{AB}^i\rangle$, Alice and Bob may identify blocks of members in each pure state $|\psi_{AB}^i\rangle$, and apply the Schmidt projection method to them, giving $np_i S(\varrho_B^i)$ maximally entangled pairs, and hence a total entanglement of distillation of $\sum_i p_i S(\varrho_B^i)$. The relative entropy of entanglement again quantifies the amount of entanglement distillable from the state $\varrho_{ABM}^{\varepsilon}$ and is given by $E_{\mathrm{RE}}(\varrho_{A:(BM)}^{\varepsilon}) = \min_{\sigma_{ABM} \in D} S(\varrho_{ABM}^{\varepsilon} || \sigma_{ABM})$. The disentangled state which minimizes the relative entropy is

$$\sigma_{ABM} = \sum_i p_i \sigma_{AB}^i \otimes |m_i\rangle\langle m_i|, \qquad (10.26)$$

where σ_{AB}^i is obtained from $|\psi_{AB}^i\rangle\langle\psi_{AB}^i|$ by deleting the off-diagonal elements in the Schmidt basis. This is the minimum because the state ϱ_{MAB} is a mixture of the orthogonal states $|m_i\rangle|\psi_{AB}^i\rangle$, and for a pure state $|\psi_{AB}^i\rangle$, the disentangled state that minimizes the relative entropy is σ_{AB}^i. The minimum relative entropy of the extended system is then

$$S(\varrho_{ABM}^{\varepsilon} || \sigma_{ABM}) = \sum_i p_i S(\varrho_B^i)$$

This relative entropy, $E_{\mathrm{RE}}(\varrho_{A:(BM)}^{\varepsilon})$, has been called the 'entanglement of projection' (by Garisto and Hardy in 1999), because the measurement on the memory projects the pure state of the full system into a particular decomposition. The minimum of $E_{\mathrm{RE}}(\varrho_{A:(BM)}^{\varepsilon})$ over all decompositions is equal to the entanglement of formation of ϱ_{AB}. However, Alice and Bob may choose to create the state ϱ_{AB} by using a decomposition with higher entanglement than the entanglement of formation. The maximum of $E_{\mathrm{RE}}(\varrho_{A:(BM)}^{\varepsilon})$ over all possible decompositions is called the "entanglement of assistance" of ϱ_{AB}. Because $E_{\mathrm{RE}}(\varrho_{A:(BM)}^{\varepsilon})$ is a relative entropy, it is invariant under local operations and nonincreasing under general operations, properties which are conditions for a good measure of entanglement. However, unlike $E_{\mathrm{RE}}(\varrho_{AB})$ and $E_{\mathrm{F}}(\varrho_{AB})$, it is not zero for completely disentangled states. In this sense, the relative entropy of entanglement $E_{\mathrm{RE}}(\varrho_{A:(BM)}^{\varepsilon})$ defines a class of entanglement measures that interpolate between the entanglement of formation and the entanglement of assistance. Note that an upper bound for the entanglement of assistance E_A can be obtained using concavity, namely $E_A(\varrho_{AB}) \leq \min[S(\varrho_A), S(\varrho_B)]$. This bound can also be obtained from the fact that the entanglement distillable from any decomposition, $E_{\mathrm{RE}}(\varrho_{A:(BM)}^{\varepsilon}) \leq E_A(\varrho_{AB})$, cannot be greater than the entanglement of the original pure state.

Note that here we are really creating a state $\varrho^{\otimes n} = \varrho \otimes \varrho \ldots \varrho$. The entanglement of formation of such a state is, strictly speaking, given by $E_{\mathrm{F}}(\varrho^{\otimes n})$, so the entanglement of formation per one is $E_{\mathrm{F}}(\varrho^{\otimes n})/n$. It is not clear at present if this is the same as $E_{\mathrm{F}}(\varrho)$ in general, that is, whether the entanglement of formation is additive. Bearing this in mind, we shall continue our discussion, whose conclusions will not depend on the validity of the assumption of additivity of the entanglement of formation.

We may also derive the relative-entropy measures that interpolate between the relative entropy of entanglement and the entanglement of formation by considering nonorthogonal measurements on the memory. First of all, the fact that the entanglement of formation is in general greater than the upper bound on the entanglement of distillation, emerges as a property of the relative entropy, namely that it cannot increase under the local operation of tracing one subsystem;

$$\begin{aligned} E_{\mathrm{F}}(\varrho_{AB}) &= \min_{\sigma_{ABM} \in \mathcal{E}D} S(\varrho_{ABM}||\sigma_{ABM}) \\ &\geq \min_{\sigma_{AB} \in \mathcal{E}D} S(\varrho_{AB}||\sigma_{AB}) \,. \end{aligned} \quad (10.27)$$

In general, the loss of the information in the memory may be regarded as a result of an imperfect classical channel. This is equivalent to Alice making a nonorthogonal measurement on the memory, and sending the result to Bob. In the most general case, $\{E_i = A_i A_i^+\}$ is a POVM performed on the memory. The decomposition ξ corresponding to this measurement is composed of mixed states, namely $\xi = \{q_i, \operatorname{tr}_M(A_i \varrho_{MAB} A_i^+)\}$, where $q_i = \operatorname{tr}(A_i \varrho_{MAB} A_i^+)$. The relative entropy of entanglement of the state ϱ_{MAB}^ξ, when ξ is a decomposition of ϱ_{AB} resulting from a nonorthogonal measurement on M, defines a class of entanglement measures that interpolate between the relative entropy of entanglement and the entanglement of formation of the state ϱ_{AB}. In the extreme case where the measurement gives no information about the state ϱ_{AB}, $E_{\mathrm{RE}}(\varrho_{A:(BM)}^\varepsilon)$ becomes the relative entropy of entanglement of the state ϱ_{AB} itself. In intermediate cases, the measurement gives partial information.

So far, I have shown that measures interpolating between the entanglement of assistance and the entanglement of formation result from making orthogonal measurements on preparations of the pure state $|\psi_{MAB}\rangle$ in different bases. They may equally well be achieved by using preparation associated with the entanglement of assistance, and making increasingly nonorthogonal measurements.

10.7 Classical information and entanglement

The loss of entanglement may be related to the loss of information in the memory. There are two stages at which distillable entanglement is lost. The first is in the conversion of the pure state $|\psi_{MAB}\rangle$ into a mixed state ϱ_{ABM}. This happens because Alice uses a *classical* channel to communicate the memory to Bob. The second is due to the loss of the memory M when taking the state ϱ_{ABM} is taken to ϱ_{AB}. The amount of information lost may be quantified by the difference in mutual information between the respective states. The mutual information is a measure of the correlation between the memory M and the system AB, giving the amount of information about AB which may be obtained from a measurement on M. The quantum mutual information between M and AB is defined as $I_\mathrm{Q}(\varrho_{M:(AB)}) = I(\varrho_M : \varrho_{AB}) = S(\varrho_M) + S(\varrho_{AB}) - S(\varrho_{MAB})$. The mutual information loss in going from the pure state $|\psi_{MAB}\rangle$ to the mixed state in eqn 10.25 is $\Delta I_\mathrm{Q} = S(\varrho_{AB})$. There is a corresponding reduction in the relative entropy of entanglement, from the entanglement of the original pure state, $E_{\mathrm{RE}}(|\psi_{(MA):B}\rangle \langle \psi_{(MA):B}|)$, to the entanglement of the mixed state, $E_{\mathrm{RE}}(\varrho_{A:(BM)}^\varepsilon)$, for all decompositions ε arising as the result of an orthogonal measurement on the memory. It is possible to prove, using the nonincrease of the relative entropy under local operations, that when the mutual-information loss is added to the relative entropy of entanglement of the mixed state $E_{\mathrm{RE}}(\varrho_{A:(BM)}^\varepsilon)$, the result is greater than the relative entropy of entanglement of the original pure state, $E_{\text{RE}}(|\psi_{(MA):B}\rangle\langle\psi_{(MA):B}|)$. The strongest case, which occurs when $E_{\text{RE}}(\varrho^\varepsilon_{A:(BM)}) = E_{\text{F}}(\varrho_{AB})$, is

$$E_{\text{RE}}(|\psi_{(MA):B}\rangle\langle\psi_{(MA):B}|) \leq E_{\text{F}}(\varrho_{AB}) + S(\varrho_{AB}) . \tag{10.28}$$

A similar result may be proved for the second loss, due to loss of the memory. Again the mutual-information loss is $\Delta I_Q = S(\varrho_{AB})$. The relative entropy of entanglement is reduced from $E_{\text{RE}}(\varrho^\varepsilon_{A:(BM)})$, for any decomposition ε resulting from an orthogonal measurement on the memory, to $E_{\text{RE}}(\varrho_{AB})$, the relative entropy of entanglement of the state ϱ_{AB} with no memory. When the mutual-information loss is added to $E_{\text{RE}}(\varrho_{AB})$, the result is greater than $E_{\text{RE}}(\varrho^\varepsilon_{A:(BM)})$. In this case, the result is strongest for $E_{\text{RE}}(\varrho^\varepsilon_{A:(BM)}) = E_A(\varrho_{AB})$:

$$E_A(\varrho_{AB}) \leq E_{\text{RE}}(\varrho_{AB}) + S(\varrho_{AB}) \tag{10.29}$$

Note that if ϱ_{AB} is a pure state then $S(\varrho_{AB}) = 0$, and equality holds. The inequalities 10.28 and 10.29 provide lower bounds on $E_{\text{F}}(\varrho_{AB})$ and $E_{\text{RE}}(\varrho_{AB})$, respectively. They are of a form typical of irreversible processes, in that restoring the information in M is not sufficient to restore the original correlation between M and AB. In particular, they express the fact that the loss of entanglement between Alice and Bob at each stage must be accompanied by an even greater reduction in the mutual information between the memory and the subsystem AB. A general result can be derived from Donald's equality. In general, for any σ and $\varrho = \sum_i p_i \varrho_i$, the following is true:

$$S(\varrho||\sigma) + \sum_i p_i S(\varrho_i||\varrho) = \sum_i p_i S(\varrho_i||\sigma) .$$

Suppose that $E(\varrho) = S(\varrho||\sigma)$. Then, since $E(\varrho_i) \leq S(\varrho_i||\sigma)$, we have the following inequality:

$$E(\varrho) + \sum_i p_i S(\varrho_i||\varrho) \geq \sum_i p_i E(\varrho_i) .$$

Thus, the loss of entanglement in $\{p_i, \varrho_i\} \to \varrho$ is bounded from above by the Holevo bound;

$$\sum_i p_i E(\varrho_i) - E(\varrho) \leq \sum_i p_i S(\varrho_i||\varrho) . \tag{10.30}$$

This is a physically pleasing property of the entanglement. It says that the amount of information lost always exceeds the lost entanglement, which indicates that entanglement stores only part of the information—the rest, of course, is stored in classical correlations.

I close this section by discussing generalizations to more than two subsystems. First of all, it is not at all clear how to perform such a generalization in the case of the entanglement of formation and of distillation. The entanglement of formation just does not have a natural generalization, and, for the entanglement of distillation, it is not clear what states should we be distilling when we have three or more parties. The relative entropy of entanglement, on the other hand, does not suffer from this problem. Its definition for N parties would be $E_{\text{RE}}(\sigma) := \min_{\varrho \in D} S(\sigma||\varrho)$, where $\varrho = \sum_i p_i \varrho_1^i \otimes \varrho_2^i \ldots \otimes \varrho_N^i$.

10.8 Entanglement and thermodynamics

Mathematically the most rigorous and sound way of speaking about the Second Law of Thermodynamics is to view it as a rule that permits us to introduce an order into the set of physical states. This order encodes information about the fact that certain states are more likely to occur than others, and are in some sense "preferred" by nature. The function that introduces this order among states is the well-known entropy, and the Second Law then implies that in any adiabatic process the entropy of the system undergoing transformation cannot decrease. Let us take a simple example to show why this law of no decrease of entropy during adiabatic processes is to be desired in the first place. Consider an ideal gas adiabatically isolated from its surrounding environment. It is clear from our experience that the gas can be heated up (by steering it, for example), thereby increasing its internal energy without doing any work and without any other changes elsewhere in the universe, since no heat exchange is allowed during the process. The entropy during this process would then increase as the gas increases its average velocity. It should at least be intuitively clear that we cannot go back to the original state of the gas and recover all this work, providing that our gas is adiabatically insulated. And this basic fact is what the Second Law sets out to encapsulate. According to Carathéodory, this simple example can be elevated to the level of a universal principle by requiring that in *any* neighborhood of *any* thermodynamic state, there are states which cannot be accessed by *adiabatic* means. The important words are the two emphasized "any"s and the word "adiabatic". If any of those are changed (i.e. the conditions are relaxed) then the law no longer leads to the ordering of states mentioned above. In fact, it turns out that Carathéodory's statement on its own is not enough to deduce the entropy. We also need the First Law of Thermodynamics and the assumption that the internal energy is a continuous function of the state variables. This dependence on the First Law can be eliminated.[9]

What is the basic idea behind Carathéodory's formulation and how does one arrive at the entropy function and the ordering of states? First of all, we have to agree that all states can be described by a set of n variables, $n-1$ of which are extensive (these are also known as the deformation coordinates, for example pressure, volume, and so on), and one of which is the internal energy (the fact that this is a complete description is assumed to be based on experimental evidence). The question is whether and how we can operationally define an entropy of a state which will tell us if that state can be converted into another state (of equal or higher entropy) or not (of lower entropy). Suppose that we choose a state Σ with coordinates x_i, U and assign it an entropy S_0. Now we choose another state $\bar{\Sigma}$ with coordinates $\bar{x}_i, \bar{U}$, and ask what the entropy of this state is. First, we deform the extensive coordinates of $\bar{\Sigma}$ so that they become identical to those of Σ, that is, $x_i = \bar{x}_i$. The assumption is that this can always be

---

[9]You can probably gather that physicists are not very fond of Carathéodory's formulation of thermodynamics, since it is very mathematical. Imagine that someone asks you what they need to do to become famous by violating the Second Law. If you were a physicist, you would say that all they need do is transfer heat from a cold to a hot body without any other effect. Nice and simple (to formulate, though not to do). Carathéodory, on the other hand, would say that what you need to do is find a state and one of its neighborhoods such that all the states in there were accessible to it adiabatically. Wow! Nevertheless, despite this cumbersome formulation, Carathéodory's approach will be very useful to us here.

done without any other changes and in an adiabatic fashion. Now, the new internal energy of this state, U', is such that we have either $U' < \bar{U}$, or $U' = \bar{U}$, $U' > \bar{U}$.

- If $U' = \bar{U}$, we say that the entropy of $\bar{\Sigma}$ is also S_0.
- If $U' < \bar{U}$, we can (irreversibly) heat $\bar{\Sigma}$ so that its internal energy equals that of Σ. The entropy of $\bar{\Sigma}$ is now defined to be $S = S_0 + (U' - \bar{U})$.
- If $U' > \bar{U}$, then we can (irreversibly) heat Σ; until its internal energy equals that of $\bar{\Sigma}$. The entropy of $\bar{\Sigma}$ is now defined to be $S = S_0 - (U' - \bar{U})$.

So, we see that the ordering of states is achieved by an entropy which is operationally defined using the (already introduced) notion of internal energy. But where exactly did Carathéodory's statement of the Second Law come in? We have in the above argument assumed implicitly that when we change a state $\bar{\sigma}$ by altering its extensive variables, we reach a unique state with a well-defined (unique) internal energy. This may not be the case in reality. Here is where Carathéodory's statement becomes important, since it guarantees that the state is in fact unique. The proof is by *reductio ad absurdum*. Suppose that there are two such states. We can then show that there exist neighborhoods of these where all states are adiabatically accessible, in contradiction with Carathéodory's statement.

Carathéodory's method is very beautiful; however, in spite of its mathematical formulation, there are still some assumptions that we have had to make *en passant*. One of those is that given any two states Σ_1 and Σ_2, there exists an adiabatic process for going either from Σ_1 to Σ_2 or the other way round. If there are states that are mutually inaccessible then the above argument fails (actually, we shall see later that this assumption can be relaxed, although not by much). The second crucial assumption is that we are relying on properties of the internal energy derived from the First Law, in particular the continuity of the internal energy, that is, if we change the extensive variables by a small amount then the internal energy changes by a small amount as well. This in turn implies that the entropy is also continuous, as it is derived from the internal energy. These additional assumptions may give us the impression that the Second Law is in some sense dependent on the First Law. But this would be completely erroneous: the Second Law is logically completely independent.

The first person to spell this out was Buchdahl, and his approach was subsequently entirely formalized by Giles. We shall now review Giles' treatment and show how this formalization achieves more universal appeal in that it can be applied to quantum mechanics to quantify entanglement. To begin with, we study a nonempty set $\mathcal{E}S$, whose elements are called **states**, on which two operations, $+$ and $\rightarrow$, are defined. States are denoted by $= a, b, c, \ldots$ A **process** is an ordered pair of states (a, b). The set of all processes will be denoted by $\mathcal{E}P$. In the following axioms, we shall omit the phrase "for all …".

Axioms 1–5.

1. The operation $+$ is associative and commutative.
2. $a \rightarrow a$.
3. $a \rightarrow b \ \& \ b \rightarrow c \Longrightarrow a \rightarrow c$.
4. $a \rightarrow b \Longleftrightarrow a + c \rightarrow b + c$.
5. $a \rightarrow b \ \& \ a \rightarrow c \Longrightarrow b \rightarrow c$ or $c \rightarrow b$.

Let us briefly discuss why these axioms describe the structure of thermodynamics. The operation "+" represents the physical operation of considering two systems together. Therefore it must naturally be associative and commutative. The operation "→" represents an adiabatic process, which is meant to convert different physical states into each other. Therefore, like any other physical process, it should naturally be reflexive and transitive, as in axioms 2 and 3. Axiom 4 is the first nonintuitive property linking the operations "+" and "→". In the forward direction, it is obvious that if state a can be converted into b then the presence of another state c should not alter this fact, that is, we can convert a and c into b and c by converting a into b and doing nothing to c. In the backward direction, however, this axiom is not completely obvious. It says that if a process is possible with the aid of another state, then in fact we do not need this state for the process. Thermodynamics deals with macroscopic systems with a large number of degrees of freedom (subsystems). It is in this "asymptotic" limit that this axiom becomes more natural (this asymptotic property will become even clearer when we talk about entanglement). Finally, axiom 5 is the key property which allows us to compare different states and processes. It says that any two states that are accessible from a third state can be accessible from each other, at least in one direction. If this was not true it would lead to states that were incomparable, as there would be no physical way of connecting them.

We need the following definitions in order to introduce the remaining two axioms. Their importance is in comparing the sizes (contents/amounts) of different physical states.

Definition. Given states a and b, we write $a \subset b$ (and say that a is contained in b) if there exists a positive integer n and a state c such that

$$na + c \to nb \text{ or } nb \to na + c .$$

This really says that the state a is smaller than b if a requires the help of another state c to be converted to or derived from b.

Definition. A state e is an **internal** state if, given any state x, there exists a positive integer n such that $x \subset ne$.

This definition serves to introduce a reference state, which is one that can contain any other physical state given sufficiently many copies of it. The concept of an internal state is necessary to give us a basic metric unit (yardstick) with which to quantify the physical content of a state in a unique way (independently of the state).

The following two axioms are meant to ensure that the comparison of states never results in an infinite quantity.

Axioms 6–7

6. There exists an internal state.
7. Given a process (a, b), if there exists a state c such that for any positive real number ϵ there exist positive integers m, n and states x, y such that $m/n < \epsilon$, $x \subset mc$, $y \subset mc$, and $na + x \to nb + y$, then $a \to b$.

Axiom 6 is obviously necessary if we are to compare the contents of different states in a unique way. Axiom 7 is the most complex axiom in the theory, although it is strongly motivated by the logic of thermodynamic reasoning. Loosely speaking, it states that if we can transform a into b with an arbitrarily small environmental influence, then this influence can be ignored. This, in some sense, introduces continuity into thermodynamic properties (and will be crucial for our manipulations of entanglement later).

The above axioms are the crux of Giles' formal theory, which captures the key idea behind the Second Law of Thermodynamics. This theory proves the existence of a unique **entropy function** on the basis of the above axioms. However, Giles' formalization is general enough to be applicable to local manipulations of quantum pure states, as we shall describe in the following. This then leads to a rigorous proof of the existence of a unique **entanglement function**.

The most intriguing feature of Giles' (as well as Carathéodory's) formulation is that there is nothing special about the entropy increasing; that is, the axioms themselves do not dictate that the entropy of a system has to increase (this is an obvious feature of Giles' axioms, but it also holds for Carathéodory's formulation). In fact, all the features of this formulation would be the same even if the entropy were to decrease (i.e. $a \to b$ if and only if $S(a) \geq S(b)$). This is one of the main reasons why Carathéodory's approach was heavily criticized by Planck. Planck maintained, completely correctly, that the law that the entropy increases has to be imposed in addition rather then be derived from the postulates which describe thermodynamic processes. This would mean that there is no physical reason why the entropy should increase rather than decrease—as far as Giles (and Carathéodory) is concerned, we could equally well be living an a world where entropy decreases. Far from being a disadvantage, we think that this is an enormous advantage of Giles' system; the reason is that we can think of a different physical situation where the axioms apply, but where, the quantity that orders the states tells us that the dynamics is such that entropy cannot increase. This happens exactly to be the case when we try to quantify bipartite entanglement on the basis of which entangled states can be converted to which other entangled states using only the local operations and classical communication.

The advantage of Giles' formulation is that by formalizing a physical theory in such a way as to divorce it completely from its physical interpretation, it is possible to derive a mathematical structure that may be useful in a completely different physical setting from the original one. The analogy between thermodynamics and quantum information that I described earlier identifies adiabatic processes with LOCC and entropy with entanglement. The aim is to order quantum states according to whether they can be converted into each other by LOCCs. It can be shown—the details are tedious and uninteresting—that Giles' framework can be applied to quantifying entanglement in a pure bipartite state. As a consequence, we can arrive at a rigorous proof of the existence of a unique measure of entanglement for pure states in the same way as a unique measure of order (entropy) is constructed in thermodynamics. If we specialize to qubits, we can then say that one bit of information (entanglement) exists in a state which can be converted into any other state by LOCC, but not vice versa. Therefore, if we have a maximally entangled state of two qubits, $(|00\rangle + |11\rangle)/\sqrt{2}$, we arrive at one bit of entanglement. Disentangled states, on the other hand, are assigned zero information (or entanglement). So, a state with higher information can be converted into a state with the same amount of information or less, but not vice versa and this is the **sole** basis for defining the concept of information. This formulation has the power that it frees us from the need to define probabilities in order to define information.

We begin by reminding ourselves of the concept of local operations aided by classical communication (LOCC) used in manipulating entangled states. As we have seen, LOCC plays a crucial role in understanding the notion of the degree of entanglement between two systems. This is so because LOCC is able to separate disentangled states from entangled states and thus introduce a **directionality** to entanglement manipulation processes: an entangled state can always be converted into a disentangled one by LOCC, but not vice versa. Therefore we can imagine that we take an entangled state as our reference point and assign to it a certain amount of entanglement. Then we apply LOCC to it and see which other states can be reached (these would consequently be assigned less entanglement).

So, how do we interpret Giles' axioms within the quantum mechanical formalism so as to describe entanglement? We can think of the states "$a, b, c, \ldots$" as quantum mechanical pure states that consist of two subsystems A and B, whose total Hilbert space is $H_A \otimes H_B$. The operation "+" is then interpreted as the direct product "$\otimes$" of two states. The arrow "$\to$" is then a special quantum mechanical process that transforms quantum states into each other but where the only operations that are allowed are of local nature.

What about mixed bipartite states? In this case it is very likely that one (or more) of the axioms is violated, hence preventing us from concluding that there is a unique measure of entanglement for mixed states. The axiom that is now difficult to prove is number 5; it is no longer clear that any two states which can locally be reached from a common state are accessible to each other at least in one direction. One way of correcting this difficulty is to change the definition of LOCC, and admit a wider class of operations such that axiom 5 becomes satisfied. If, however, we extend LOCC, we shall involve some nonlocal operations as well, and so the resulting quantity (which would be unique as a result) would no longer quantify entanglement. Therefore, we have decided to abandon this direction. Suppose that, instead, we relax axiom 5 to make it weaker. We may require the following:

5'. If $a \to b$ and $a \to c$, then there exist d and e such that $b \subset d$ and $c \subset e$ and either $d \to e$ or $e \to d$.

This is correct, as we can think of the states d and e as purifications of b and c (and they then contain b and c), which are convertible to each other one way or the other. Thus, if replace axiom 5 by 5', then even mixed states satisfy all the axioms. The problem now is whether these extended axioms lead to a unique measure of entanglement.[10] This is an open question and it is a possible avenue to explore when addressing the issues of multiparty entanglement.

I shall now use the knowledge we have gained about classical and quantum correlations to describe quantum computation. It will be seen, not at all surprisingly I suppose, that quantum correlations play a role in the speedup of certain quantum algorithms.

---

[10] Entanglement may intrinsically require more than one measure for quantification. The problem of quantifying entanglement is a bit like the problem economists face in trying to put a price on everything. Some things may require other quantifiers apart from the price.

10.9 Summary

It is difficult to argue for a unique measure of entanglement for mixed states, although some natural rules that all measures have to satisfy can be written down. All measures should produce zero for separable states and should not increase under LOCC. It may also be natural to require continuity, but beyond this it is very difficult to argue for more requirements. These requirements on their own will not produce a unique measure of entanglement.

The three most prominent measures are the entanglement of distillation, the entanglement of formation, and the relative entropy. All of these satisfy the above requirements and have a strong operational meaning.

Furthermore, we have seen that the relative entropy of entanglement of the state ϱ_{AB} depends only on the density matrix ϱ_{AB}, and gives an upper bound on the entanglement of distillation. The other measures of entanglement, which are given by relative entropies of an extended system, all depend on how the information in a memory is used or on how the density matrix is decomposed. There are numerous decompositions of any bipartite mixed state into a set of states ϱ_i with probabilities p_i. The average entanglement of the states in each decomposition is given by the relative entropy of entanglement of the system, extended by a memory whose orthogonal states are classically correlated with the states of the decomposition. This correlation records which state ϱ_i each member of an ensemble of mixed states $\varrho_{AB}^{\otimes n}$ is in. It is available to the parties involved in the formation of the mixed state, but is not accessible to parties carrying out distillation. When the classical information is fully available, different decompositions give rise to different amounts of distillable entanglement, the highest being the entanglement of assistance and the lowest being the entanglement of formation. If the access to the classical record is reduced, the amount of distillable entanglement is reduced. In the limit where no information is available, the upper bound on the distillable entanglement is given by the relative entropy of entanglement of the state ϱ_{AB} itself, without extension by a classical memory.

Part III

Quantum Computation

11
Quantum algorithms

As computers get faster and faster, the size of the circuitry imprinted onto silicon chips decreases. Moore's law, based upon Moore's observation that over the past 50 years there has been an exponential growth in the number of transistors per integrated circuit, says that this trend will continue through the twenty-first century until eventually the size of the circuitry is so small that its behavior is governed by the laws of quantum mechanics. Such a computer, whose computations would be fully quantum mechanical, is called a quantum computer.

Any computational task such as addition, multiplication, displaying graphics, or updating databases is performed by a computer according to an algorithm—an abstract set of instructions. A classical algorithm, performed in a world governed by classical mechanics, can be written as a sequence of logical operations performed on a bit string. Quantum computers would accomplish tasks by performing quantum algorithms. A quantum algorithm is a sequence of unitary evolutions carried out on a quantum string made up of qubits, which can exist as a superposition of classical strings. An advantage of quantum algorithms over classical algorithms is that some tasks can be carried out much faster on a quantum computer by taking advantage of the fact that quantum algorithms can process many classical strings in superposition. The difficult part for a quantum computer is the final measurement of the output, which always has to select only one of the strings in the superposition.

11.1 Computational complexity

The computational complexity of an algorithm is the number of elementary steps (whatever these are defined to be) it takes (Papadimitriou (1995) is an excellent introduction into this subject). Consider a simple problem, that of counting the number of squares in a box, shown in Fig. 11.1. A naive algorithm might count each square individually. A much faster algorithm might just count the number of squares along one of the sides of the box and use this to calculate the actual number of squares. If the box has sides of length n, we say that the naive algorithm takes n^2 steps to count all the squares, whereas the more efficient algorithm takes only n steps. We can say that the more efficient algorithm is more efficient than the naive algorithm by a factor of n.

A well-known problem is 3-colorability. As shown in Fig. 11.2, the problem is, given a set of points connected by edges, to decide whether the points can be colored using only three colors so that no two points connected by an edge are colored in the same color. One way to solve this problem is to try out all the different combinations

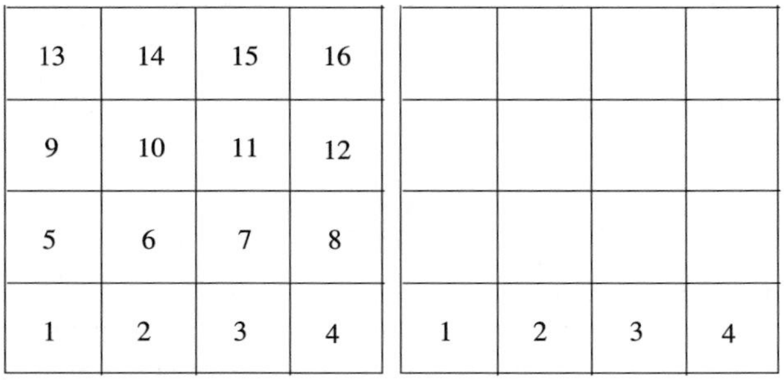

132 *Quantum algorithms*

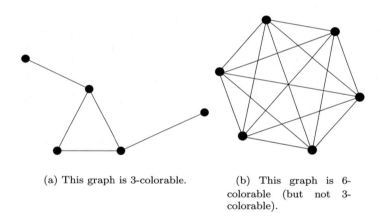

(a) A naive algorithm for counting the squares in a box counts them individually.

(b) A more efficient algorithm just counts the number of squares along one side and uses this value to calculate the total number of squares in the box.

Fig. 11.1 Two ways of counting the number of squares in a box.

(a) This graph is 3-colorable.

(b) This graph is 6-colorable (but not 3-colorable).

Fig. 11.2 The 3-colorability problem is to find whether a set of points connected by edges (lines) can be colored in three colors so that no two points connected by an edge have the same color.

of colors. If there are n points, then this algorithm would take 3^n steps to test each combination.

The run-times of the above algorithms are compared in Table 11.1 for some values of n. The 3-colorability algorithm takes many more steps for larger values of n than either of the other two algorithms. For a graph with $n = 1000$ points, the 3-colorability algorithm takes so many steps that it is unfeasible to carry it out on a

n	Efficient squares	Naive Squares	3-Colorability
1	1	1	1
10	10	100	59 049
100	100	10 000	5.15×10^{47}
1000	1000	1 000 000	1.3×10^{477}

Table 11.1

state-of-the-art computer (to get an idea of how large 1.3×10^{477} is, it is estimated that there are fewer than 10^{100} atoms in the universe[1]). Algorithms which take an exponential number of steps (e.g. the 3-colorability algorithm) grow much faster than algorithms which take a polynomial number of steps (e.g. either of the two square-counting algorithms). Exponential-time algorithms are often called "unfeasible" or "hard", whereas polynomial-time algorithms are often called "feasible" or "easy". It is unknown whether there exists a polynomial-time classical or quantum algorithm for solving 3-colorability. This problem lies within a special class of problems called NP-complete problems.[2] The Clay Mathematics Institute offers a $1 million prize for showing whether there is a classical polynomial-time algorithm that solves the 3-colorability problem.

11.2 Deutsch's algorithm

Deutsch's algorithm is perhaps the simplest example of a quantum algorithm which outperforms a classical algorithm. The aim of Deutsch's algorithm is to decide whether a coin is fair or biased. The coin is represented by a function f which can take as input 0 or 1. $f(0)$ represents one side of the coin, and $f(1)$ the other. If the coin is fair, then the two sides of the coin are different, that is, $f(0) \neq f(1)$. If the coin is biased, both sides of the coin are the same, that is, $f(0) = f(1)$. Using a classical algorithm, both sides of the coin are examined to determine whether the coin is fair or biased using two evaluations of the function f. Deutsch's algorithm looks at both sides of the coin in superposition and determines whether the coin is fair or biased in just one evaluation of f.

To use f in a quantum algorithm, we first design a unitary operation which can evaluate f. One operation that evaluates f is

$$X_f = |f(0)\rangle\langle 0| + |f(1)\rangle\langle 1| \,. \tag{11.1}$$

However, when f is biased, X_f is not unitary. Nevertheless, an auxiliary system can be used to design an operation for evaluating f which is unitary. Let U_f be defined by

$$U_f|xy\rangle = |x\rangle|y \oplus f(x)\rangle \,, \tag{11.2}$$

---

[1] It is of course very difficult to estimate the number of atoms in the universe. We don't even know its boundaries or its full structure.

[2] NP-complete problems are a special class of difficult problems. NP stands for "nondeterministic polynomial", meaning that if we are given the solution to the problem then we can verify it in polynomial time (but without the solution, these are generally hard to solve). The NP-complete problems are a subclass of the NP problems such that if one of them were solved the whole class NP would become polynomial, i.e. easy. See Garey and Johnson (1979) for a detailed exposition of NP-complete problems.

134 Quantum algorithms

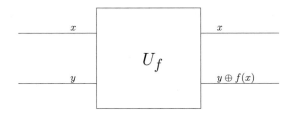

Fig. 11.3 U_f written as a gate in a circuit diagram. The two lines on the *left* represent two input wires, one for each qubit. The two lines on the *right* represent the output of the two qubits after U_f is applied.

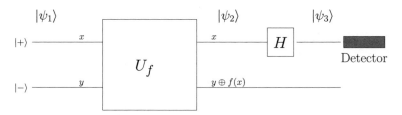

Fig. 11.4 A diagram of Deutsch's algorithm. The initial input is $|\psi_0\rangle = |01\rangle$. A Hadamard transformation is applied to each qubit to prepare a superposition $|\psi_1\rangle = (H \otimes H)|\psi_0\rangle$. The superposition $|\psi_1\rangle$ is sent through the U_f gate (only once) to obtain $|\psi_2\rangle$. Another Hadamard operation is applied to $|\psi_2\rangle$ to obtain $|\psi_3\rangle$, the second qubit of which can be measured to decide whether f is fair or biased.

where $\oplus$ denotes addition modulo 2. In other words,

$$U_f = |0f(0)\rangle\langle 00| + |0f(1)\rangle\langle 01| + |0\overline{f(0)}\rangle\langle 10| + |1\overline{f(1)}\rangle\langle 11|, \tag{11.3}$$

where $\overline{y}$ denotes the NOT operation, that is, $\overline{0} = 1$, and $\overline{1} = 0$. U_f is commonly written as a gate in a circuit diagram as shown in Fig. 11.3.

The entire circuit for Deutsch's algorithm is shown in Fig. 11.4. The input to Deutsch's algorithm is prepared in the state

$$|\psi_1\rangle = |+-\rangle, \tag{11.4}$$

which can be prepared from $|01\rangle$ using two Hadamard gates. We now evaluate f using U_f:

$$|\psi_2\rangle = U_f|\psi_1\rangle \tag{11.5}$$
$$= U_f\left(\frac{1}{2}(|0\rangle + |1\rangle)(|0\rangle - |1\rangle)\right) \tag{11.6}$$
$$= \frac{1}{2}(U_f|00\rangle - U_f|01\rangle + U_f|10\rangle - U_f|11\rangle) \tag{11.7}$$
$$= \frac{1}{2}(|0f(0)\rangle - |0\overline{f(0)}\rangle + |1f(1)\rangle - |1\overline{f(1)}\rangle) \tag{11.8}$$
$$= \frac{1}{2}|0\rangle(|f(0)\rangle - |\overline{f(0)}\rangle) + |1\rangle(|f(1)\rangle - |\overline{f(1)}\rangle). \tag{11.9}$$

If f is fair, $f(1) = \overline{f(0)}$. If f is biased, $f(1) = f(0)$. Depending on whether f is fair or biased, we obtain two values for $|\psi_2\rangle$:

$$|\psi_2^{\text{fair}}\rangle = \frac{1}{2}|0\rangle(|f(0)\rangle - |\overline{f(0)}\rangle) + |1\rangle(|f(0)\rangle - |\overline{f(0)}\rangle) \tag{11.10}$$

$$= |+\rangle \left(\frac{|f(0)\rangle - |\overline{f(0)}\rangle}{\sqrt{2}} \right), \tag{11.11}$$

$$|\psi_2^{\text{biased}}\rangle = \frac{1}{2}|0\rangle(|f(0)\rangle - |\overline{f(0)}\rangle) + |1\rangle(|\overline{f(0)}\rangle - |f(0)\rangle) \tag{11.12}$$

$$= |-\rangle \left(\frac{|f(0)\rangle - |\overline{f(0)}\rangle}{\sqrt{2}} \right). \tag{11.13}$$

Applying the Hadamard transformation to the first qubit yields

$$|\psi_3^{\text{fair}}\rangle = |0\rangle \left(\frac{|f(0)\rangle - |\overline{f(0)}\rangle}{\sqrt{2}} \right), \tag{11.14}$$

$$|\psi_3^{\text{biased}}\rangle = |1\rangle \left(\frac{|f(0)\rangle - |\overline{f(0)}\rangle}{\sqrt{2}} \right). \tag{11.15}$$

If the first qubit is now measured as $|0\rangle$ then f is fair, and if the first qubit is measured as $|1\rangle$ then f is biased.

Let us now rephrase this in terms of phase shifts to emphasize its underlying identity with the Mach–Zehnder interferometer experiment described earlier. The transformation of the two registers between $|\phi_1\rangle$ and $|\phi_3\rangle$ is the following:

$$|x\rangle|-\rangle \Rightarrow e^{i\pi f(x)}|x\rangle|-\rangle,$$

where $x = 0, 1$. Thus, the first qubit is like a photon in the interferometer, receiving a conditional phase shift depending on its state (0 or 1). The second qubit is there just to implement the phase shift quantum mechanically.[3] This quantum computation, although extremely simple, contains all the main features of successful quantum algorithms: it can be shown that all quantum computations are just more complicated variations of Deutsch's problem.

11.2.1 Deutsch's algorithm and the Holevo bound

The input to a quantum computer is rarely a pure state. There is generally noise present from all sorts of different sources, which results in the computer being in a mixed state. We shall see in Chapter 13 how to address errors in a systematic way through error correction, but let's see before that how the efficiency of quantum computation is reduced by mixing.

The Holevo bound bounds the quantum speedup when we try to input states mixed together with $|+\rangle|+\rangle$ into U_f. Suppose that the input is of the form $|-\rangle|-\rangle$. A biased function would then lead to the state $|-\rangle|-\rangle$ and a fair function would lead to $|+\rangle|-\rangle$.

---
[3] You can think of it as the state of the beamsplitters.

So $|+\rangle$ and $|-\rangle$ are equally good as input states of the first qubit and both lead to quantum speedup. An equal mixture of them, on the other hand, is not. In that case the output would be an equal mixture $|+\rangle\langle+|+|-\rangle\langle-|$ no matter whether $f(0) = f(1)$ or $f(0) \neq f(1)$, that is, the two possibilities would be indistinguishable. Thus for a quantum algorithm to work well, we need the first register to be highly correlated with the two different types of functions. So, if the output state of the first qubit ϱ_1 indicates that we have a biased function and ϱ_2 that we have a fair function, then the efficiency of Deutsch's algorithm depends on how well we can distinguish the two states ϱ_1 and ϱ_2. This is given by the Holevo bound

$$\chi = S(\varrho) - \frac{1}{2}(S(\varrho_1) + S(\varrho_2)),$$

where $\varrho = (\varrho_1 + \varrho_2)/2$. Thus if $\varrho_1 = \varrho_2$, then $\chi = 0$ and the quantum algorithm has no speedup over the classical algorithm. At the other extreme, if ϱ_1 and ϱ_2 are pure and orthogonal, then $\chi = 1$ and the computation gives the right result in one step (this is the ideal case considered earlier). In between these two extremes lie all other computations, with varying degrees of efficiency as quantified by the Holevo bound. The key to understanding the efficiency of Deutsch's algorithm is therefore through the mixedness of the first register. If the initial state has an entropy S_0, then the final Holevo bound is

$$S(\varrho) - S_0 .$$

So the more mixed the first qubit, the less efficient the computation. This is a general conclusion for any algorithm, not just Deutsch's. We now look at Oracles and then other important algorithms.

11.3 Oracles

Oracles are a formal way of describing how many time steps an algorithm takes. An Oracle is a function f, and the number of steps that an algorithm takes is the number of times f is evaluated. At the beginning of this chapter, the Oracle for counting a square. For the 3-colorability problem, the Oracle we used was checking the graph to check that no two vertices connected by an edge had the same color. A more realistic Oracle would be to check whether the two points connected by an edge were the same color, as this is a much simpler operation. There are at most n^2 edges in a graph with n points, so in terms of this Oracle, the runtime is at most $n^2 3^n$, which is still exponential.

In the last section, the Oracle f represented checking the value of one side of the coin. We compared the classical and quantum algorithms by considering a unitary transformation U_f out of f and comparing the number of times f and U_f are evaluated. This is the key in comparing quantum and classical algorithms. If f is a classical function then we can create a unitary quantum Oracle as follows:

$$O_f(|x\rangle) = (-1)^{f(x)}|x\rangle . \tag{11.16}$$

We then say that the time complexity of a quantum algorithm using O_f is the number of times the unitary operation O_f is executed. This Oracle is equivalent[4] to one

---

[4] By "equivalent" I mean that the two Oracles can simulate each other polynomially.

where the value of the function is recorded into another register, as used in Deutsch's algorithm $|x\rangle|0\rangle \to |x\rangle|f(x)\rangle$.

11.4 Grover's search algorithm

Suppose you have a long, unordered list of N names and addresses of people. To find the person X who lives at 1 Quantum Avenue requires reading $N/2$ addresses on average, and N addresses in the worst case when 1 Quantum Avenue is the last address read. Grover's algorithm is a quantum algorithm which, by reading the addresses in superposition, can find who lives at 1 Quantum Avenue in just $\sqrt{N}$ steps.

This search problem can be formulated more formally as follows: given an integer N and a function f such that

$$\begin{aligned} f(i) &= 0 \quad \text{if } i \neq x \text{ and } 0 \leq i < N, \\ f(i) &= 1 \quad \text{if } i = x, \end{aligned} \tag{11.17}$$

find the value of x. If 1 Quantum Avenue is the xth address, then $f(i)$ returns 1 only if the ith address is the one we are looking for. The output of the algorithm is x, which we can then use to find who lives there by quickly looking up the xth entry in the list. This problem is much more general than finding the person who lives at a particular address and can be applied to searching any unordered database.

Here is how a quantum search achieves its remarkable speedup. Initially the input is prepared in an equal superposition of all states,

$$|\psi\rangle = |+\rangle^{\otimes n} = \sum_i \frac{1}{\sqrt{N}} |i\rangle . \tag{11.18}$$

The algorithm works by applying an Oracle O_x followed by a unitary transformation U and repeating UO_x for T steps, where T is to be found. After t steps, the input is transformed to

$$|\psi^t\rangle = (UO_x)^t |\psi\rangle . \tag{11.19}$$

Let us define the Oracle and then define U. The Oracle flips the phase of x:

$$\begin{aligned} O|x\rangle &= -|x\rangle, & (11.20) \\ O|i\rangle &= |i\rangle \text{ if } i \neq x . & (11.21) \end{aligned}$$

We can write the Oracle as an operation as follow:

$$\begin{aligned} O &= -|x\rangle\langle x| + \sum_{i \neq x} |i\rangle\langle i| & (11.22) \\ &= \sum_i |i\rangle\langle i| - 2|x\rangle\langle x| & (11.23) \\ &= |\psi\rangle\langle\psi| - 2|x\rangle\langle x|, & (11.24) \end{aligned}$$

where $|\psi\rangle$ is defined in eqn 11.18. We also use another phase-shifting gate. Let $O_{\delta 0}$ be a conditional phase shift acting on 0:

$$O_{\delta 0}|0\ldots 0\rangle = -|0\ldots 0\rangle \quad (11.25)$$
$$O_{\delta 0}|i\rangle = |i\rangle \text{ if } i \neq 0 \quad (11.26)$$

This can be written likewise as

$$O_{\delta 0} = \sum_i (|i\rangle\langle i|) - 2|0\rangle\langle 0| = I - 2|0\ldots 0\rangle\langle 0\ldots 0| \quad (11.27)$$

Now we define U. U is a unitary operation made up of n Hadamard gates, followed by a conditional phase shift acting on 0, followed by n Hadamard gates:

$$U = H^{\otimes n} O_{\delta 0} H^{\otimes n} . \quad (11.28)$$

To find out what U does, we remember that the Hadamard operation can be written as

$$H = |0\rangle\langle +| + |1\rangle\langle -| , \quad (11.29)$$

so that

$$\langle 0\ldots 0|H^{\otimes n} = \langle 0\ldots 0|(|0\rangle\langle +| + |1\rangle\langle -|)^{\otimes n} \quad (11.30)$$
$$= \langle +\ldots +| . \quad (11.31)$$

We use this equation to multiply out U:

$$U = H^{\otimes n}(I - 2|0\ldots 0\rangle\langle 0\ldots 0|)H^{\otimes n} \quad (11.32)$$
$$= H^{\otimes n}(H^{\otimes n} - 2|0\ldots 0\rangle\langle +\ldots +|) \quad (11.33)$$

The Hadamard operation is self-inverse, and so $H^{\otimes n} H^{\otimes n} = I$ and the coefficients of $H^{\otimes n}$ are real; therefore $H^{\otimes n}|0\ldots 0\rangle = \langle 0\ldots 0|H^{\otimes n} = \langle +\ldots +|$. Plugging these two formulas in, we get

$$U = I - 2|+\ldots +\rangle\langle +\ldots +| \quad (11.34)$$
$$= I - 2|\psi\rangle\langle\psi| , \quad (11.35)$$

and we can rewrite U as

$$U = 2|\psi\rangle\langle\psi| - I , \quad (11.36)$$

which is the same up to a global phase.

A beautiful analysis of Grover's search algorithm[5] can be obtained from considering $|x\rangle$ as the x-axis and $\sum_{i \neq x} |i\rangle/\sqrt{N-1}$ as the y-axis in a two-dimensional plane as shown in Fig. 11.5. We define

$$|y\rangle = \frac{1}{\sqrt{N-1}} \sum_{i \neq x} |i\rangle = |\psi\rangle - |x\rangle , \quad (11.37)$$

and we define an angle θ so that

---
[5] This analysis is due to Richard Jozsa.

Grover's search algorithm

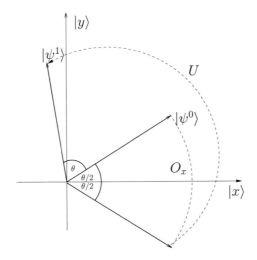

Fig. 11.5 The first step of Grover's algorithm. The Oracle O_x is a reflection in the x-axis, and U is a reflection in $|\psi\rangle$. Together, U and O_x form a rotation anticlockwise by an angle of θ.

$$\cos\left(\frac{\theta}{2}\right) = \sqrt{\frac{N-1}{N}}. \tag{11.38}$$

The initial state of the system $|\psi\rangle$ can be written in terms of $|x\rangle$, $|y\rangle$, and θ as

$$|\psi\rangle = \cos(\theta/2)|y\rangle + \sin(\theta/2)|x\rangle. \tag{11.39}$$

Figure 11.5 shows the geometry of the algorithm. Now, for any α, the effect of the Oracle is

$$O_x(\cos(\alpha)|y\rangle + \sin(\alpha)) = \cos(\alpha)|y\rangle - \sin(\alpha)|x\rangle, \tag{11.40}$$

which is a reflection in the x-axis. What is the geometrical interpretation of U? Well,

$$\begin{align}
U|x\rangle &= (2|\psi\rangle\langle\psi| - I)|x\rangle \tag{11.41}\\
&= 2\sin(\theta/2)|\psi\rangle - |x\rangle \tag{11.42}\\
&= (1 - 2\sin^2(\theta/2))|x\rangle - 2\cos(\theta/2)\sin(\theta/2)|y\rangle \tag{11.43}\\
&= -\cos(\theta)|x\rangle + \sin(\theta)|y\rangle, \tag{11.44}
\end{align}$$

$$\begin{align}
U|y\rangle &= (I - 2|\psi\rangle\langle\psi|)|y\rangle \tag{11.45}\\
&= 2\cos(\theta/2)|\psi\rangle - |y\rangle \tag{11.46}\\
&= -2\cos(\theta/2)\sin(\theta/2)|x\rangle + (1 - 2\cos^2(\theta/2))|y\rangle \tag{11.47}\\
&= \sin(\theta)|x\rangle + \cos(\theta)|y\rangle, \tag{11.48}
\end{align}$$

and so U is a reflection in $|\psi\rangle$. Together, the effect of U and O_x is:

$$\begin{align}
UO_x|x\rangle &= \cos(\theta)|x\rangle - \sin(\theta)|y\rangle, \tag{11.49}\\
UO_x|y\rangle &= \sin(\theta)|x\rangle + \cos(\theta)|y\rangle, \tag{11.50}
\end{align}$$

which is a rotation by an angle θ. Thus the effect of UO_x on the initial state $|\psi^0\rangle$ is

$$|\psi^1\rangle = UO_x|\psi^0\rangle \qquad (11.51)$$
$$= UO_x(\cos(\theta/2)|y\rangle + \sin(\theta/2)|x\rangle) \qquad (11.52)$$
$$= \cos(3\theta/2)|y\rangle + \sin(3\theta/2)|x\rangle . \qquad (11.53)$$

And after t rotations, the state of the system is rotated by an angle $t\theta$ and is in the state

$$|\psi^t\rangle = (UO_x)^t|\psi^0\rangle \qquad (11.54)$$
$$= \cos\left(\frac{(2t+1)\theta}{2}\right)|y\rangle + \sin\left(\frac{(2t+1)\theta}{2}\right)|x\rangle . \qquad (11.55)$$

If we allow the algorithm to go on forever, it will keep rotating round and round. We want to stop it after T steps when it is in the state $|x\rangle$, in which case

$$\sin\left(\frac{(2T+1)\theta}{2}\right) = 1 . \qquad (11.56)$$

If we express the angles in radians, this happens when

$$\frac{(2T+1)\theta}{2} = \pi . \qquad (11.57)$$

When N is large, $\sin(\theta)$ is small, so $\theta \approx \sin(\theta) = 1/\sqrt{N}$ and

$$\frac{2T+1}{2}\frac{1}{\sqrt{N}} \approx \pi . \qquad (11.58)$$

So the number of iterations or time steps required to solve the algorithm is

$$T = 2\pi\sqrt{N} - 1 . \qquad (11.59)$$

However, $2\pi\sqrt{N} - 1$ is not a whole number. But we can choose the nearest integer and still obtain a good probability of success. As N becomes large, T grows at a rate less than N and Grover's algorithm outperforms any classical algorithm.

This wraps up the formal part of the analysis of the search algorithm. We shall have much more to say about it in the next chapter in relation to quantum measurement. Next we turn to another famous quantum algorithm.

11.5 Quantum factorization

Prime numbers are of fundamental importance in number theory and have fascinated mathematicians for many years. Prime numbers are the building blocks of arithmetic, yet their distribution seems to be quite random. There are many seemingly simple problems to do with prime numbers which have baffled many mathematicians for centuries. One such problem is to find whether there exists a fast algorithm for factorization. The aim of factorization is, given an integer n, to find the prime numbers p_1, ... p_k such that $n = p_1 \ldots p_k$. It is an open problem whether there exists a polynomial time classical algorithm (polynomial in terms of the number of bits $\log(n)$) to find the factorization of n.

Public key cryptography is a widely used form of cryptography, for example, it is used on the Internet for making credit card transactions. Alice thinks of two large prime numbers p and q and makes $n = pq$ publicly available. If Bob wants to talk to Alice, he takes his message and performs some arithmetic involving the message and n so that the message can be read with knowledge of p and q, that is, by Alice. The idea behind ~~cryptography~~ this type of cryptography is that an eavesdropper Eve cannot find a fast (polynomial time) algorithm to compute p and q from n. The numbers p and q are chosen sufficiently high so that it is impractical to factorize n without a fast algorithm. If a fast algorithm were to be found, then public key cryptography would no longer be a secure means of communication. The applications of breaking public key cryptography include being able to read people's credit card details as they make transactions over the Internet.

In 1994, Shor developed a quantum algorithm which can factorize efficiently. This algorithm (published before Grover's) was the first useful algorithm for a quantum computer which outperforms a classical computer. It fueled the rapid expansion of interest in quantum computation in the late 1990s.[6] Here we give the main ideas behind the algorithm. An interested reader can find the details in almost any textbook on quantum computation.

In this section, we show that factorization can be performed by phase estimation, then we describe the quantum Fourier transform, which is used in phase estimation, and then we describe phase estimation itself.

11.5.1 Factorization

We can factor an integer N by estimating the eigenvalues of a unitary operation. Let U_a be a unitary operation defined by

$$U_a|x\rangle = |ax \bmod N\rangle. \tag{11.60}$$

There is then some value r (called the order of a) such that $a^r \bmod(N) = 1$ which implies that $U_a^r = I$. By considering the spectral decomposition of U_a, we can show that this implies that each eigenvalue λ_k of U_a is an rth root of unity:

$$\lambda_k = e^{2\pi i k/r} \tag{11.61}$$

The reader can check (by noting that $U_a|a^j\rangle = |a^{j+1}\rangle$ or otherwise) that the corresponding eigenvector is

$$|e_k\rangle = \frac{1}{\sqrt{r}} \sum_{j=0}^{r-1} e^{-2\pi i k j/r} |a^j \bmod(N)\rangle. \tag{11.62}$$

Suppose we had a procedure for finding the eigenvalues of U_a. We could run this procedure several times to obtain m values k'_j/r'_j of k/r (where k'_j and r'_j share no common factor), from which we could guess r with a high probability of success. Let

---

[6]I joined the field in the first half of 1995, straight after Shor's publication.

$k_j/r = k'_j/r'_j$. Then the lowest common multiple R of the r'_j's is equal to r if and only if there is no integer $s > 1$ which divides all the k_j's. The probability that an integer s divides some k_j is $1/s$, and hence the probability that s divides $k_1, \ldots, k_m$ is $1/s^m$. The probability that there is some integer $s > 2$ that divides $k_1, \ldots, k_m$ is at most $\sum_{s=2}^{\infty} 1/s^m < 3 \times 2^{-m}$ which is very small. Therefore, with a high probability of success, we can guess that $R = r$.

If we can find the eigenvalues of U_a, we can now find the order r of $a \bmod N$. That is, we have a quantum algorithm for finding the order of an integer $a \bmod N$. It is wellknown in number theory that we can use an order finding algorithm to efficiently factor an integer. Let's see how...

Suppose we are trying to factor some integer N and we have found the order r of some a randomly chosen between 1 and N. With a high probability of success we can find a factor of N. We can check whether a and N are coprime using Euclid's algorithm which finds the greatest common divisor of two integers. If a and N share a common divisor (greater than 1), then we have found a factor of N and are done. Otherwise, the order of a is r, that is, $a^r \bmod N = 1$. If r is even, then N divides $a^r + 1 = (a^{r/2} - 1)(a^{r/2} + 1)$. Now if N divides $a^{r/2} - 1$ then $a^{r/2} \bmod N = 1$, and so the order of a would be less than r. Suppose also that N does not divide $(a^{r/2} + 1)$. Then N shares common factors with $(a^{r/2} + 1)$ and $(a^{r/2} - 1)$, which can be found using Euclid's algorithm. We can check whether this is the case by checking that $a^{r/2} \neq -1 \bmod N$, in which case this algorithm succeeds, otherwise it fails. It can be shown that the probability that the algorithm succeeds is at least $1/2$. If we repeat the algorithm t times, the probability of failure is $1/2^t$, which is exponentially small.

11.5.2 The quantum Fourier transform

The quantum Fourier transform is used in many quantum algorithms, including phase estimation (which is how we estimate the eigenvalues of U_a). The Hadamard operation is in fact a quantum Fourier transform on one qubit. We first define the quantum Fourier transform, and then show that it can be implemented efficiently.

The discrete Fourier transform is widely used in applied mathematics and engineering. If the discrete Fourier transform acts on a vector as follows,

$$\begin{pmatrix} y_0 \\ y_1 \\ \vdots \\ y_{N-1} \end{pmatrix} = \begin{pmatrix} & & \\ & F.T. & \\ & & \end{pmatrix} \begin{pmatrix} x_0 \\ x_1 \\ \vdots \\ x_{N-1} \end{pmatrix}, \tag{11.63}$$

then

$$y_k = \frac{1}{\sqrt{N}} \sum_{j,k=0}^{N-1} x_j e^{2\pi i j k/N}. \tag{11.64}$$

The discrete Fourier transform can be generalized to the quantum Fourier transform under which a vector $|j\rangle$ is transformed to

$$|j\rangle \to \frac{1}{\sqrt{N}} \sum_{j=0}^{N-1} e^{2\pi i j k/N} |k\rangle. \tag{11.65}$$

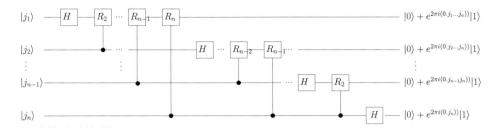

Fig. 11.6 The quantum Fourier transform on n qubits. A black dot together with R_k denotes the control bit of CR_k. The breakdown of the quantum Fourier transform into elementary unitary gates shows that the quantum Fourier transform is unitary itself.

A superposition under a quantum Fourier transform is transformed to:

$$\sum_{j=0}^{N-1} \alpha_j |j\rangle \to \sum_{k=0}^{N-1} \alpha_j e^{2\pi i \frac{jk}{N}} |k\rangle \qquad (11.66)$$

The discrete Fourier transform is hard to calculate on a classical computer, however the quantum Fourier transform is divided up easily into a circuit. For simplicity, let us assume that $N = 2^n$ for some integer n. Let $J = J_1 \ldots J_n$ be a binary expansion of an integer J so that $J = 2^{n-1} J_1 + \ldots + 2^0 J_n$. We can do the same for the part of a number after the decimal. Let $j = 0.j_l \ldots j_m$ be an binary expansion of j so that $j = j_l/2 + \ldots + j_m/2^{m-l+1}$.

The action of the quantum Fourier transform can be written as follows:

$$|j_0 \ldots j_{n-1}\rangle \to \frac{1}{\sqrt{2^n}} (|0\rangle + e^{2\pi i (0.j_n)}) \otimes (|0\rangle + e^{2\pi i (0.j_{n-1} j_n)})$$
$$\otimes \ldots \otimes (|0\rangle + e^{2\pi i (0.j_1 j_2 \ldots j_n)}) \qquad (11.67)$$

This formula gives us the circuit for the quantum Fourier transform shown in Fig. 11.6. Each $|j_i\rangle$ is a qubit. We can write the action of the Hadamard transform on $|j_i\rangle$ as

$$H|j_i\rangle = \frac{1}{\sqrt{2}} (|0\rangle + e^{2\pi i (0.j_i)} |1\rangle) . \qquad (11.68)$$

We also use another rotation transformation,

$$R_k = \begin{pmatrix} 1 & 0 \\ 0 & e^{2\pi i/k} \end{pmatrix} . \qquad (11.69)$$

However, we control R_k on the basis of another value:

$$CR_k |0\rangle |j\rangle = |0\rangle |j\rangle , \qquad (11.70)$$
$$CR_k |1\rangle |j\rangle = |0\rangle R_k |j\rangle . \qquad (11.71)$$

How the circuit works is best shown by a diagram. The reader can check that Fig. 11.7 shows the implementation of the quantum Fourier transform on three qubits. The general quantum Fourier transform is shown in Fig. 11.6.

144 *Quantum algorithms*

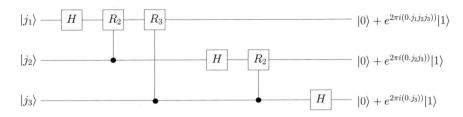

Fig. 11.7 The quantum Fourier transform on three qubits.

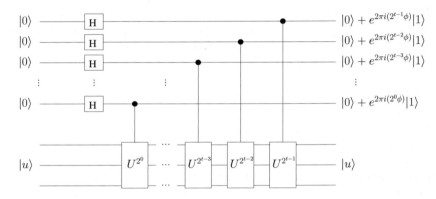

Fig. 11.8 The first part of the phase estimation algorithm. The algorithm is finished by applying the inverse of the quantum Fourier transform to the first registers.

11.5.3 Phase estimation

The aim of phase estimation is, given an eigenvector $|e\rangle$ of a unitary transformation U, to find the corresponding eigenvalue λ. We can apply the phase estimation operation U_a used for factorization above to do this because the sum of an equal superposition of U_a's eigenvectors $|e_k\rangle$ is equal to $|1\rangle$, and applying phase estimation to $|1\rangle$ produces one of the eigenvalues λ_k with uniform probability. The circuit for the phase estimation is shown in Fig. 11.8.

Let's imagine for simplicity that ϕ has a decimal expansion of t bits as follows:

$$\phi = 0.\phi_1 \ldots \phi_t . \tag{11.72}$$

We can then rewrite the output of Fig. 11.8 as

$$\frac{1}{\sqrt{2^t}}(|0\rangle + e^{2\pi i(0.\phi^t)}|1\rangle) \otimes (|0\rangle + e^{2\pi i(0.\phi^{t-1}\phi^t)}|1\rangle) \otimes (|0\rangle + e^{2\pi i(0.\phi^1 \ldots \phi^t)}|1\rangle) \tag{11.73}$$

which looks amazingly like the eqn 11.67, which expands the quantum Fourier transform. If we now apply the quantum Fourier transform in reverse (we invert it by applying the circuit backwards), then we obtain $|\phi_1 \ldots \phi_t\rangle$. This phase estimation algorithm can be implemented for factorization efficiently by use of repeated squaring—$a^{2x} = (a^x)^2$—in order to calculate U^{2^t} quickly. However in practice ϕ is likely to have a longer decimal representation than $0.\phi_1 \ldots \phi_t$, but it is close, so we have a good probability of success.

We can think of all of the quantum algorithms presented above as phase estimations simply because the Oracle operation encodes the functional value into the phase of the computational register, and this register then has to be manipulated in order to obtain the value of the phase and hence the function.

11.6 Summary

All quantum algorithms exploit the fact that a superposition of different input states is processed simultaneously by a quantum computer and this allows us to speed certain operations up with respect to a classical computer which can only process one input at a time. The three algorithms we presented were:

- Deutsch's algorithm, whose task was to determine if a binary function of a single binary variable was constant (fair) or varying (biased).
- Grover's algorithm, which searches an unsorted database of N elements in $\sqrt{N}$ steps.
- Shor's algorithm, which achieves factorization of an N-bit number in polynomial time — approximately N^3 steps.

Each of these clearly demonstrates that quantum computation is more efficient than classical.

We do not really know the ultimate efficiency of quantum computers, but already this discussion shows that quantum computation is a worthwhile adventure. Can quantum algorithms help us understand something about quantum mechanics? The next chapter, where we talk about quantum measurements, shows that the answer clearly is "yes".

12
Entanglement, computation and quantum measurements

I would now like to introduce a method that combines what we have learned about quantum entanglement with what we have learned about quantum algorithms. Namely we shall put some bounds on the efficiency (speedup) of quantum computations by using the fact that there is a limit to how quickly entanglement can be generated with some operations. This method will truly exploit the "many-worlds" nature of quantum superpositions. We shall not only be solving one problem using a superposition of qubits, but shall also be solving a superposition of problems![1] This view will give us a considerable advantage in evaluating the efficiency of some quantum tasks.

Quite generally, we can think of computations in the following way. The initial state of one of the registers of the quantum computer is prepared in an equal superposition of all possible states. One or more of these states represents the solution to our problem, but of course, we do not know which—otherwise, we would just prepare the solution to start with! The computation then consists of using another register to query the first one in order to find the right solution.

What happens if every state in the initial superposition in the first register—which we shall call the memory register—is a solution, but to a different problem? Well, the second register—which we shall call the computational register—could then solve all these problems simultaneously by means of a suitable interaction! So we would start with a disentangled state of problems

$$(|p_1\rangle + |p_2\rangle + |p_3\rangle + \ldots |p_n\rangle)|0\rangle \qquad (12.1)$$

and, through a sequence of computational steps, would end with the state:

$$|p_1\rangle|s_1\rangle + |p_2\rangle|s_2\rangle + |p_3\rangle|s_3\rangle + \ldots |p_n\rangle|s_n\rangle \, , \qquad (12.2)$$

which is a maximally entangled state between the two registers of problems and solutions.

And so what would be the efficiency of solving all n problems at the same time? (Of course, in practice we would have to make a measurement in the end and only one of the problems would end up being solved, as it were, but since this picture is

---

[1] If you think that quantum mechanics is great because it allows us to solve some problems by exploiting superpositions, then you should remember that problems also become more complicated in quantum mechanics because they can be in superpositions as well!

only a calculational tool we don't really care about this point). The efficiency would be the same as that of creating the $\log n$ bits of entanglement between the registers, which start disentangled and end up maximally entangled. Therefore, in this way, the question of the computational efficiency reduces to the efficiency of entanglement generation (with whatever gates we are allowed to use).

I shall now show how this logic can be applied to demonstrating that a quantum search cannot be performed faster than the square root of the time of the corresponding classical search (which is still very much faster!). This is a very fundamental result and has a wide range of implications.

12.1 Optimization of searches using entanglement

The approach I shall follow here is due to Ambainis.[2] We start with a register which contains the database element that we are looking for.

Any general quantum algorithm has to include a certain number of queries of the memory register (this is necessitated by the fact that the transformation performed on the computational register has to depend on the problem at hand, encoded in $|i\rangle_M$). These queries can be considered to be implemented by a black box[3], into which the states of both the memory and the computational registers are fed. The number of such queries needed in a given quantum algorithm gives the black-box complexity of that algorithm and is a lower bound on the complexity of the whole algorithm.

The black-box approach is a simplification of the problem of looking at the complexity of an algorithm. A black box allows us to perform a certain computation without knowing its exact details. It is possible that the physical implementation of a particular black box may prove to be difficult. So when we estimate the complexity of an algorithm by counting the number of applications of a black box, we have to bear in mind that there might an additional component of complexity arising from the physical implementation.

In general we have a function $f : \{0,1\}^n \rightarrow \{0,1\}$ (so the function maps n-bit values to either 0 or 1). Quantum algorithms, such as Grover's algorithm (for database search), can be expressed in this form (in the case of the database search, all the values of f are 0 apart from one value which is equal to 1; the task is to find this value). The black box is assumed to be able to perform the transformation $|x\rangle|y\rangle \rightarrow |x\rangle|f(x) \oplus y\rangle$, just as in Deutsch's algorithm. We have the freedom to represent this black-box transformation as a phase flip which is equivalent in power (up to a constant factor);

$$|x\rangle|y\rangle \rightarrow (-1)^{f(x) \oplus y}|x\rangle|y\rangle.$$

Ambainis showed that if the memory register is prepared initially in the superposition $\sum_i^N |i\rangle_M$, then, in a search algorithm, $O(\sqrt{N})$ queries are needed to completely

---

[2]Who is himself a computer scientist. Computer scientists seem to be much more at ease with quantum parallelism than are physicists. This may be because physicists still do not fully accept the possibility of macroscopic superpositions (despite so many macroscopic manifestations of superpositions, such as superconductivity, superfluidity, and Bose condensation).

[3]A black box is another way of describing an Oracle. We assume that the black box implements some function and it is the black box that is used to query the function.

entangle it with the computational register. This gives a lower bound on the number of queries in a search algorithm. We shall follow his proof here.

Any search algorithm (whether quantum or classical, and irrespective of its explicit form) will have to find a match for the state $|i\rangle_M$ of the register M among the states $|j\rangle_C$ of the register C and associate a marker with the state that matches. (Here, $|j\rangle_C$ is a complete orthonormal basis for the C register.) The most general way of doing such a query in the quantum case is the black-box unitary transformation

$$U_B|i\rangle_M|j\rangle_C = (-1)^{\delta_{ij}}|i\rangle_M|j\rangle_C. \tag{12.3}$$

Any other unitary transformation that performs a query match the states of the registers M and C could be constructed from the above type of query. Note that the black box is able to recognize if a value in the register C is the same as the solution, but is unable to explicitly provide that solution for us. To illustrate this, imagine that Socrates goes to visit the all-knowing ancient Greek oracle (the black box) who is only able to answer "yes" or "no". Suppose that Socrates wants to know who the wisest person in the world is. He would then have to ask something like "Is Plato the wisest person in the world?" and would not be able to ask directly "Who is the wisest person in the world?" This "yes–no" approach is typical of any black-box analysis. The advantage of using this black box quantum mechanically is that we can query all the individual elements of the superposition simultaneously. Although we can identify the solution in one step quantum mechanically, further computations are required to amplify the right solution so that the subsequent measurement is more likely to reveal it.

The derivation of the number of steps required to reach the solution will go as follows. We shall show that it takes $\sqrt{N}$ applications of the box to maximally entangle the two registers. Since the registers start disentangled, each is in a pure state. The more they get entangled, the more mixed they become and it is this mixedness (specifically, the reduction in the size of the off-diagonal elements) that will tell us how quickly the two registers become fully entangled.

First we want to show that the sum of the off-diagonal elements in the register M cannot decrease by more than $1\sqrt{N}$ in one step. Let us first write down the state after n steps of the black-box iteration.

Step n:

$$\sum_{ij} \alpha_{ij}|i\rangle|j\rangle \rightarrow \varrho = \sum_{\substack{ij \\ kl}} \alpha_{ij}\alpha_{kl}^*|ij\rangle\langle kl|. \tag{12.4}$$

The reduced state of the memory register can easily be obtained to be:

$$\text{tr}_A \Rightarrow \sum_{ijl} \alpha_{ij}\alpha_{il}^*|j\rangle\langle l| \leftarrow \text{Density operator of memory.} \tag{12.5}$$

So the off-diagonal elements of the density matrix are

$$c(j,l) = \sum_i \alpha_{ij}\alpha_{il}^*,$$

where $j \neq l$. In the next step, we have the following:

Step n+1:

$$\varrho' = \sum_{ij} \alpha_{ij}(-1)^{\varrho_{ij}}|i\rangle|j\rangle(\langle\Psi|)$$

$$\text{tr}_A \Rightarrow \sum_{ijl} \alpha_{ij}\alpha_{il}^*(-1)^{\delta_{ij}+\delta_{il}}|j\rangle\langle l|$$

is the density operator of the memory after one application of the black box.

Now, we compute the difference in the off-diagonal elements between nth and first step when $n = 1$:

$$\sum_{j \neq l}|c(j,l)| - |c'(j,l)| \leq \sum_{j \neq l}|c(j,l) - c'(j,l)|$$

$$\leq \sum_{i,j \neq l}|\alpha_{ij}\alpha_{il}^*|(1-(-1)^{\delta_{ij}+\delta_{il}})$$

$$= 2\sum_{i \neq l}|\alpha_{ii}\alpha_{il}^*| + 2\sum_{i \neq j}|\alpha_{ij}\alpha_{ij}^*|$$

$$\leq 4\sum_{i \neq l}|\alpha_{ii}||\alpha_{il}|$$

$$\sim N^2 \tfrac{1}{N} \times \tfrac{1}{\sqrt{N}}$$

$$\leq 4\sqrt{N} \ .$$

One application of the black box therefore reduces the sum of the off-diagonal elements by roughly $\sqrt{N}$. The initial sum is about N, since we start from an equal superposition of all possibilities. The final sum should be zero, when we reach maximum entanglement. We need to go from $\sim N$ to 0, and the number of steps that are required is at most $\sqrt{N}$.

Therefore, the efficiency of any search algorithm is at best of the order of $O(\sqrt{N})$. In fact, this efficiency can be achieved using Grover's search algorithm as we saw in the previous chapter. Entanglement thus offers a very general tool for proving how efficient various algorithms are in terms of the number of black boxes applications.[4]

12.2 Model for quantum measurement

The above way of viewing the efficiency of quantum computation through entanglement is very closely related to viewing a measurement as the establishment of entanglement between the system and the apparatus. I shall first describe a one-qubit system, but shall then generalize this to connect it to von Neumann's treatment of measurement of continuous variables (Neumann 1955) such as the position of a particle.[5]

---

[4] Again, here we have been searching over all databases at the same time. The final state is one where the computer is in a superposition of states, each matching the solution for one database. This fits perfectly the many-worlds view of quantum mechanics. In this world, I solve one problem, but in another world, i.e. branch of the superposition, I solve another problem, and so on ad infinitum.

[5] Von Neumann's work on quantum measurement is perhaps the most profound statement that has been made about quantum measurements, in spite of having been made 80 years ago!

The measurement of one qubit can be represented in the following way:

$$(a|0\rangle + b|1\rangle)|m\rangle$$
$$\to a|0\rangle|m_0\rangle + b|1\rangle|m_1\rangle ,$$

where the overlap $\langle m_0|m_1\rangle = \xi$ determines the strength of the measurement. If the states of the apparatus are orthogonal at the end, that is, $\xi = 0$, the measurement is the strongest possible, and if the overlap is small, the measurement is weak. So how entangled the apparatus and the system become determines the strength of the measurement. All this, and much more, we saw in Chapter 3 when we discussed generalized measurements, POVMs, and CP-maps.

Grover's search algorithm, which works on any number of elements N, is therefore equivalent to a discretized measurement of the type

$$(|1\rangle + |2\rangle + \ldots |N\rangle)(|1\rangle + |2\rangle + \ldots |N\rangle) \xrightarrow[\text{Oracle}]{\text{Evolution}}$$

$$\begin{aligned} & |1\rangle(-|1\rangle + |2\rangle + \ldots |N\rangle) \\ + & |2\rangle(|1\rangle - |2\rangle + \ldots |N\rangle) \\ & \vdots \\ + & |N\rangle(-|1\rangle + |2\rangle + \ldots - |N\rangle) \end{aligned} \quad \leftarrow \text{Nonzero reduced entropies}$$

Therefore, the speed of measurement is determined by the speed of entanglement generation by black boxes, and this of course depends on the nature of the black box. Some black boxes have greater entanglement-generating power than others, but they may also be more difficult to design and implement in practice.

I now show that quantum computation is formally identical to a quantum measurement as described by von Neumann. The analysis will be performed in the most general continuous case, of which everything that we have analyzed above is just a special case. Suppose that we have a system S (described by a continuous variable x) and an apparatus A (described by a continuous variable y interacting via a Hamiltonian $H = xp$, where p is the momentum of A); we shall assume that $\hbar = 1$. Suppose in addition that the initial state of the total system is

$$|\Psi(0)\rangle = \int_x \phi(x)|x\rangle \, dx \otimes \eta(y)|y\rangle \, dy$$

in an uncorrelated state. The action of the above Hamiltonian then transforms the state into an entangled state. In order to calculate this transformation, it will be beneficial to introduce the (continuous) Fourier transform

$$F_y : |y\rangle \to \int e^{-iyp}|p\rangle \, dp ,$$

which takes us from the position space of A into the momentum space of A. This is important because we know the effect of the Hamiltonian in the momentum basis. Now, the action of the unitary transformation generated by H is

$$\begin{aligned}|\Psi(t)\rangle &= e^{-ixpt}|\Psi(0)\rangle \\ &= F_y e^{-ixpt} F_y |\Psi(0)\rangle \\ &= \int_x \int_y \phi(x)\eta(y-xy)|x\rangle|y\rangle \, dx \, dy \, ,\end{aligned}$$

and we see that S and A are now correlated in x and y. This means that by measuring A, we can obtain some information about the state of S. The mutual information $I_{AS} = H(x) + H(y) - H(x,y)$ can be shown to satisfy (Everett 1973)

$$I_{AS} \geq \ln t \, ,$$

that is, it grows at a rate faster than the logarithm of the time that has passed during the measurement. This gives us a lower bound on exactly how quickly correlations can be established between the system and the apparatus.

I now show the detailed calculation of the effect of the measurement Hamiltonian. Let us define

$$\xi(p) := F_y\{\eta(y)\}$$

The evolution then proceeds as follows:

$$\begin{aligned}|\Psi(t)\rangle &= e^{-xpt} \int_x \phi(x)|x\rangle \, dx \otimes \eta(y)|y\rangle \, dy \\ &= e^{-xpt} \int_x \phi(x) \int_p \left\{ \int_y \eta(y) e^{-iyp} \, dy \right\} |x\rangle|p\rangle \, dx \, dp \\ &= e^{-xpt} \int_x \int_p \phi(x)\xi(p)|x\rangle|p\rangle \, dx \, dp \\ &= \int_x \phi(x) \int_y \left\{ \int_p \xi(p) e^{-ixpt} e^{iyp} \, dp \right\} |x\rangle|y\rangle \, dx \, dy \\ &= \int_x \int_y \phi(x)\eta(y-xy)|x\rangle|y\rangle \, dx \, dy \, .\end{aligned}$$

This result has the same formal structure as that of the quantum algorithms presented before: a Fourier transform, followed by a conditional phase shift, followed by another Fourier transform (compare Deutsch's, Shor's, and Grover's algorithms). Therefore we can see that how efficiently we can measure something is the same as how efficiently we can compute something, both of which ultimately depend on how quickly we can establish correlations.

12.3 Correlations and quantum measurement

Any measurement can be modeled as an establishment of correlations between two random variables: one random variable represents values of a quantity pertaining to the system to be measured, while the other random variable represents states of the apparatus used to measure the system. It is by looking at the states of the apparatus and discriminating between them that we infer the states of the system. Looking at the apparatus, of course, is another measurement process itself, which correlates our mental state (presumably another random variable) with the states of the apparatus, so that indirectly we become correlated with the system as well. It is at this point that we can say that we have gained a certain amount of information about the system. This description of the measurement process is true in both classical and quantum physics. The difference between the two lies in the way we represent states of systems and the way we represent their mutual interaction and evolution. Classically, the physical states of an n-dimensional system are vectors in a real n dimensional vector space whose elements are various probabilities of occupation for the states. The evolution of a classical system is in general some stochastic map acting on this vector space. Quantum mechanically, on the other hand, states are in general represented using density matrices, and the evolution is a completely positive, trace-preserving transformation acting on these matrices. Using this representation, classical physics becomes a limiting case of quantum mechanics when the density matrices are strictly diagonal in one and the same fixed basis, and the completely positive map then becomes a stochastic map.[6] Because of this fact, it is enough to analyze the properties of quantum systems and quantum evolutions, and all the results are automatically applicable to classical physics when we restrict ourselves to diagonal density operators only.

We shall now analyze a quantum measurement when the apparatus is "fuzzy", that is, it is initially in a mixed state. We first remind ourselves of the measures of entangled and total correlation introduced earlier. In classical information theory, the Shannon entropy $H(X) \equiv H(p) = -\sum_i p_i \log p_i$ is used to quantify the information in a random variable X that contains states x_i with probabilities p_i. In the quantum context, the results of a projective measurement $\{E_y\}$ on a state represented by a density matrix ϱ comprise a probability distribution $p_y = \text{tr}(E_y \varrho)$. Von Neumann showed that the lowest entropy of any of these probability distributions generated from the state ϱ was achieved by the probability distribution composed of the eigenvalues of the state, $\lambda = \{\lambda_i\}$. This probability distribution would arise from a projective measurement onto the state's eigenvectors. The von Neumann entropy is then given by $S(\varrho) = -\text{tr}(\varrho \log \varrho) = H(\lambda)$. The quantum relative entropy of a state ϱ with respect to another state σ is defined as $S(\varrho||\sigma) = -S(\varrho) - \text{tr}(\varrho \log \sigma)$. The joint entropy $S(\varrho_{AB})$ for a composite system ϱ_{AB} consisting of two subsystems A and B is given by $S(\varrho_{AB}) = -\text{tr}(\varrho_{AB} \log \varrho_{AB})$ and the von Neumann mutual information between the two subsystems is defined as $I(\varrho_{AB}) = S(\varrho_A) + S(\varrho_B) - S(\varrho_{AB})$. The mutual information is the relative entropy between ϱ_{AB} and $\varrho_A \otimes \varrho_B$. The mutual information is used to measure the total correlation between the two subsystems of a bipartite quantum system. The entanglement of a bipartite quantum state ϱ_{AB} may be measured by how distinguishable it is from the "nearest" separable state, as measured by the relative entropy. The relative entropy of entanglement, defined as

$$E_{\text{RE}}(\varrho_{AB}) = \min_{\sigma_{AB} \in D} S(\varrho_{AB}||\sigma_{AB}) \tag{12.6}$$

has been shown to be a useful measure of entanglement (here D is the set of all separable, or disentangled states). Note that $E_{\text{RE}}(\varrho_{AB}) \leq I(\varrho_{AB})$, by the definition of $E_{\text{RE}}(\varrho_{AB})$, since the mutual information is also the relative entropy between ϱ_{AB} and

---

[6]This is not transparent to many physicists...

a completely disentangled state. There are many other ways of measuring the entanglement of a bipartite quantum state, but they can all be unified under the formalism of the relative entropy as we have seen earlier. Another advantage of the relative entropy is that it can be generalized to any number of subsystems, a property that will be very useful in understanding the measurement process when the environment is also present. I shall drop the subscript "RE" used to denote the relative entropy of entanglement as this is the only measure that will be used throughout. As a general comment, we stress that all the measures used here are entropic in nature, which means that they are generally attainable only asymptotically. The advantage of using entropic measures is that our results will be universally valid, although they will almost always be overestimates in the finite-case scenario.

The correlations in a state ϱ_{AB} can be split into two parts, the quantum and the classical part. The classical part can be seen as the amount of information about one subsystem, say A, that can be obtained by performing a measurement on the other subsystem, B. The resulting measure is the difference between the initial and the residual entropy:

$$C_B(\varrho_{AB}) = \max_{B_i^\dagger B_i} S(\varrho_A) - \sum_i p_i S(\varrho_A^i) \tag{12.7}$$

where $B_i^\dagger B_i$ is a positive operator-valued measure performed on subsystem B, and $\varrho_A^i = \mathrm{tr}_B(B_i \varrho_{AB} B_i^\dagger)/\mathrm{tr}_{AB}(B_i \varrho_{AB} B_i^\dagger)$ is the remaining state of A after the outcome i the outcome has been obtained from B. Alternatively, $C_A(\varrho_{AB}) = \max_{A_i^\dagger A_i} S(\varrho_B) - \sum_i p_i S(\varrho_B^i)$ if the measurement is performed on subsystem A instead of on B. Clearly $C_A(\varrho_{AB}) = C_B(\varrho_{AB})$ for all states ϱ_{AB} such that $S(\varrho_A) = S(\varrho_B)$ (e.g. pure states). It remains an open question whether this is true in general (but this will not affect our analysis of measurement, as the apparatus is always measured to infer the state of the system and never the other way round). This measure is a natural generalization of the classical mutual information, which is the difference in uncertainty about subsystem B (or A) before and after a measurement on the correlated subsystem A (or B, respectively). Note the similarity of the definition to the Holevo bound, which measures the capacity of quantum states for classical communication. The following example provides an illustration of this and will be key to our discussion of quantum measurement. Consider a bipartite separable state of the form $\varrho_{AB} = \sum_i p_i |i\rangle\langle i|_A \otimes \varrho_B^i$, where the $\{|i\rangle\}$ are orthonormal states of subsystem A. Clearly the entanglement of this state is zero. The best measurement that Alice can make to gain information about Bob's subsystem is a projective measurement onto the states $\{|i\rangle\}$ of subsystem A. Therefore the classical correlations are described by

$$C_A(\varrho_{AB}) = S(\varrho_B) - \sum_i p_i S(\varrho_B^i), \tag{12.8}$$

which is, for this state, equal to the mutual information $I(\varrho_{AB})$ between the two parts ϱ_A and ϱ_B. This is to be expected, since there are no entangled correlations and so the total measure of the correlation between A and B should be equal to the measure of classical correlation. This measure of classical correlations has other important properties, such as $C(\varrho_{AB}) = 0$ if and only if $\varrho_{AB} = \varrho_A \otimes \varrho_B$; it is also invariant under local unitary transformations and nonincreasing under any general local operations.

Let us now introduce the general framework for a quantum measurement. We have a system in the state $|\Psi\rangle = \sum_i a_i |i\rangle$, and an apparatus in the state $\varrho = \sum_i r_i |r_i\rangle\langle r_i|$ in the eigenbasis. The purpose of a measurement is to correlate the system with the apparatus so that we can extract information about the state $|j\rangle$ of the system. In a perfect measurement, we can unambiguously identify the state of the system by looking at the apparatus. Therefore, when the system is in the state $|j\rangle$, we would like the apparatus to be in the state ϱ_j, such that $\varrho_i \varrho_j = 0$, that is, different states of the apparatus lie in orthogonal subspaces and can be discriminated with unit efficiency. If this condition is not fulfilled, which is frequently the case, then the measurement is imperfect and the amount of information obtained is not maximal (this is what defines an "imperfect measurement"). We shall now compute the amount of information gained in general and show that it is more appropriately identified with the classical rather than the quantum correlations between the system and the apparatus. Suppose that the measurement transformation is given by a unitary operator U, acting on both the system and the apparatus, such that

$$U(\varrho \otimes |i\rangle\langle j|)U^\dagger = \varrho_{ij} \otimes |i\rangle\langle j| \,, \tag{12.9}$$

where we assume that the measurement transformation acts such that a state $|r_k\rangle|l\rangle$ of the apparatus and the system is transformed into a state $|\tilde{r}_{kl}\rangle|l\rangle$, such that the states of the apparatus corresponding to different states of the system are orthogonal $\langle \tilde{r}_{ij}|\tilde{r}_{ik}\rangle = \delta_{jk}$. This particular interaction has been chosen so that in the special case of the pure apparatus we obtain von Neumman's[7] analysis. We see that the measurement is such that the new state of the apparatus depends on the state of the system. This is exactly how correlations between the two are established. The initial state is then transformed into

$$\varrho_f = \sum_{ij} a_i a_j^* \varrho_{ij} \otimes |i\rangle\langle j| = \sum_i |a_i|^2 \varrho_{ii} \otimes |i\rangle\langle i| + \sum_{i \neq j} a_i a_j^* \varrho_{ij} \otimes |i\rangle\langle j| \,. \tag{12.10}$$

The first term on the right-hand side indicates how much information this measurement carries. We shall now measure the apparatus and try to distinguish the states ϱ_{ii} to the best of our ability. Once we have confirmed that the apparatus is in the state ϱ_{jj}, then we can infer that the system is in the state $|j\rangle$. The amount of information about the state of the apparatus (and hence the state of the system), I_m, is given by the Holevo bound:

$$I_m = S\left(\sum_i |a_i|^2 \varrho_{ii}\right) - \sum_i |a_i|^2 .S(\varrho_{ii}) \tag{12.11}$$

As we have seen, this quantity is also equal to the amount of classical correlations between the system and the apparatus in the state $\varrho'_f = \sum_i |a_i|^2 \varrho_{ii} \otimes |i\rangle\langle i|$, which is, in this case, the same as the von Neumann mutual information between the two. Note that this state is only classically correlated and there is no entanglement involved. The amount of entanglement in the state ϱ_f, on the other hand, will in general be nonzero.

---

[7] And Everett's.

This may be difficult to calculate. However, we can provide lower and upper bounds. The lower bound on the entanglement between the system and the apparatus is

$$E(\varrho_f) \geq S\left(\sum_i |a_i|^2 \varrho_{ii}\right) - S(\varrho_f) = S\left(\sum_i |a_i|^2 \varrho_{ii}\right) - S(\varrho) = I_m . \qquad (12.12)$$

(Note that $S(\varrho) = S(\varrho_{ii})$ here for all i, by the definition of the measurement interaction). Therefore, the entanglement between the system and the apparatus is greater than or equal to the amount of classical correlations between the two which quantifies the amount of information that the measurement carries. This shows that the information in a quantum measurement is correctly identified with the classical correlations between the apparatus and the system rather than with the entanglement or the mutual information between the two in the final state ϱ_f. Only in the limiting case of the pure apparatus do we have the result that the amount of information in the measurement is equal to the entanglement, which becomes the same as the amount of classical correlations, and in this case the sum of the quantum and classical correlations is then equal to the mutual information in the state. We stress that the amount of information gained in a single measurement will in general be less than this quantity, which can usually only be reached in the asymptotic limit.

We can recast this relationship in the form of an "uncertainty relation" between the initial mixedness of the apparatus and the amount of information gained. From the fact that $I_m = S\left(\sum_i |a_i|^2 \varrho_{ii}\right) - S(\varrho)$, we have that

$$I_m + S(\varrho) = S\left(\sum_i |a_i|^2 \varrho_{ii}\right) \leq \log N , \qquad (12.13)$$

where N is the dimension of the apparatus. Thus we see that the sum of the initial mixedness of the apparatus and the amount of information that the measurement obtains is always smaller than a given fixed value: the larger $S(\varrho)$, the smaller I_m. When ϱ is maximally mixed (and therefore $S(\varrho) = \log N$), no information can be extracted from the measurement. Note that this relation is different from the usual "information versus disturbance" law in a quantum measurement and from the usual entropic uncertainty relations of incompatible observables. Every measurement that extracts information from a quantum system also disturbs the state, and without that disturbance there would be no information gain possible. The initial state of the system in the above scenario was $\sum_i a_i |i\rangle$, and the final state is a mixture of the form $\sum_i |a_i|^2 |i\rangle\langle i|$. The disturbance to the state can be measured as a distance between the final and the initial state. We choose the relative entropy to quantify this difference. So, when the information gained in the measurement is given by I_m, the disturbance is

$$D = S\left(|\Psi\rangle\langle\Psi| \middle\| \sum_i |a_i|^2 |i\rangle\langle i|\right) = -\sum_i |a_i|^2 \log |a_i|^2 , \qquad (12.14)$$

which is the same as the maximum amount of information obtainable from this measurement. Therefore, the measurement described above always maximally disturbs the state, and the reason why this does not lead to the maximum information gain is because the state of the apparatus is mixed. The system could be disturbed less by adjusting the overlap between the states of the apparatus $|\tilde{r}_{ij}\rangle$, so that they are not orthogonal to each other. In general, we can require that $\langle \tilde{r}_{ij}|\tilde{r}_{ik}\rangle = a_{jk}$, such that $|a_{jk}| < 1$. We shall not treat this case here: it is mathematically more demanding, but does not illuminate the issue of measurement any better. Note also that a question may be raised as to why we consider the interaction between the apparatus and the system to be unitary and not of a more general kind. The reason is that any such interaction can be represented by a unitary transformation, and our analysis then applies again (although the resulting effective measurement would, in general, be less efficient than one performed unitarily).

In order to show that some form of entanglement is still important (albeit not that between the system and the apparatus), we now revisit the same measurement scenario, but from the "higher Hilbert space perspective". This is done by adding the environment to the apparatus so that the joint state, $|\Psi_{EA}\rangle$, is pure. We briefly note that our treatment differs from the usual "environment induced collapse" and decoherence. In our case, the environment is not there to cause the disappearance of entanglement between the system and the apparatus, but to purify the initially mixed state of the apparatus. We do not have any state reduction or collapse, but just different ways in which we can express the classical correlations between the system and the apparatus. The measurement transformation is now given by

$$|\Psi_{EA}\rangle \otimes \sum_i a_i |i\rangle \longrightarrow \sum_i a_i |\Psi_{EA}^i\rangle |i\rangle , \quad (12.15)$$

where $|\Psi_{EA}\rangle = \sum_i \sqrt{r_i} |e_i\rangle |r_i\rangle$ and $|e_i\rangle$ is an orthonormal basis for the states of the environment. We can see that when the environment is traced out, the state of the apparatus is equal to ϱ. Now, the measurement implements a unitary transformation so that each of the states of the apparatus changes according to which state of the system it interacts with. Therefore we can see that the ith state of the environment and the apparatus after the interaction is given by $|\Psi_{EA}^i\rangle = \sum_j \sqrt{r_j} |e_j\rangle |\tilde{r}_{ji}\rangle$. To make a link with the first picture of the measurement, we trace out the environment to obtain

$$\varrho_{A'S'} = \sum_{ij} a_i a_j^* \left(\sum_k \langle e_k|\Psi_{EA}^i\rangle \langle \Psi_{EA}^j|e_k\rangle \right) \otimes |i\rangle \langle j| , \quad (12.16)$$

and thus the quantity in parentheses can be identified with $\varrho_{ij} = \sum_k r_k |\tilde{r}_{ki}\rangle \langle \tilde{r}_{kj}|$. Therefore, since we have no access to the environment, our task is to discriminate the states ϱ_{ii} and therefore identify the corresponding states $|i\rangle$ of the system and this was done in the previous analysis. If, on the other hand, we had access to the environment, the measurement could be perfect.

We first apply entropic considerations to the "environment–apparatus–system" tripartite state. The initial and the final entropy of the environment are the same, as its state remains unchanged, and this value is the same as the initial entropy of the apparatus, $S(\varrho)$. As we have seen, this is an important quantity, as it determines how much information can be extracted from a measurement: the more mixed the initial state of the apparatus, the less information can be extracted. If the initial state is maximally mixed (it might be, for instance, a thermal state with an arbitrarily high temperature), then there can be no information gain during the measurement. The initial entropy of the apparatus is also equal to the entropy of the system and the apparatus after the measurement, $S(\varrho_{A'S'}) \equiv S(\varrho_f)$, as well as to the amount of entanglement between the environment and the system and the apparatus together, $E_{E:(A'S')}$, after the measurement. The entanglement and the mutual information between the environment and the apparatus after the measurement are always less than or equal to their values before the measurement (since the system becomes correlated with the apparatus during the measurement).

We should mention, finally, that our example is somewhat simplified in that the environment will not, in reality, be passive throughout the process. It will instead interact with both the system and the apparatus making the measurement even less effective, although all the above results still apply.

12.4 The ultimate limits of computation: the Bekenstein bound

Given a computer enclosed in a sphere of radius R and having available a total amount of energy E, what is the amount of information that it can store, and how quickly can this information be processed? The Holevo bound gives us the ultimate answer. The amount of information that can be written into this volume is bounded from the above by the entropy, that is, the number of distinguishable states that this volume can support. I shall now use a simple, informal argument to obtain this ultimate bound (see the appendix of (Tipler, 1994)); a rigorous derivation was first given by Bekenstein. The bound on the energy implies a bound on the momentum, and the total number of states in the phase space is

$$N = \frac{PR}{\Delta P \Delta R} \leq \frac{P\,R}{\hbar},$$

where the inequality follows from the Heisenberg uncertainty relations $\Delta P\, \Delta R \geq \hbar$ which limits the size of the smallest volume in the phase space to $\hbar$ in each of the three spatial directions, as in Fig. 12.1. From relativity, we have that for any particle, its momentum P is less than or equal to E/c, so that

$$S \leq \ln N \leq N \leq \frac{E\,R}{c\,\hbar},$$

which is known as the Bekenstein bound. In reality this inequality will most likely be a huge overestimate, but it is important to know that no matter how we encode information, we cannot perform better than what is given by our most accurate present theory—quantum mechanics. As an example, consider a nucleus of Hydrogen—according to the above result it can encode about 100 bits of information (I have assumed that $E = mc^2$ and that $R = 10^{-15}$ m). At present, NMR quantum computation achieves "only" one bit per nucleus (and not per nucleon!)—spin "up" and spin "down" are the two states. Here, it seems that we currently operate way below the allowed capacity.

From the Bekenstein bound we can derive a bound on the efficiency (speed) of information processing. Again my derivation will be loose, and a much more careful calculation confirms what I shall present. All the bits in the volume V cannot be

158 *Entanglement, computation and quantum measurements*

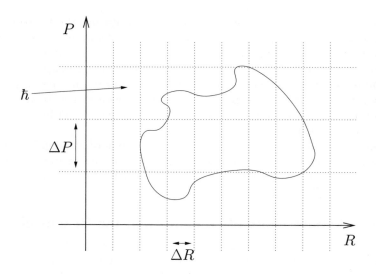

Fig. 12.1 $\hbar$ limits the size of the smallest volume in phase space.

processed faster than the time it takes light to travel across $V = (4/3)\pi R^3$, which is $2R/c$. This gives

$$\frac{dS}{dt} \leq \frac{E}{2\hbar}.$$

A Hydrogen nucleus can process about 10^{24} bits per second, which is again in sharp contrast with NMR quantum computation where a NOT gate takes roughly a few milliseconds, leading to a maximum of 10^3 bits per second.

The Bekenstein bound shows that there is a potentially great number of underused degrees of freedom in any physical system. This provides hope that large-scale quantum computation will be an experimentally realizable goal in the future. At present, there are a number of different practical implementations of quantum computation, but none of them can store and manipulate more than 10 qubits at a time. The above calculation, however, does not take into account the influence of the environment on computation nor the experimental precision. I have not touched at all on the practical possibility of building a quantum computer. This is partly for reasons of space, partly because it would spoil the flow of the exposition, and partly because there are already a number of excellent reviews of this subject. It is generally acknowledged that the difficulties in building a quantum computer are only of practical nature and there are no fundamental limits that prohibit such a device. In any case, we can see that there is a great deal of currently unused potential in physical systems with which to store and encode information. As our level of technology improves, we shall find more and more ways of getting closer to the Bekenstein bound.

12.5 Summary

Quantum measurement and quantum computation can be thought of as analogous. During a quantum measurement, one quantum system (the apparatus) interacts with another quantum system in order to make a measurement on it. This establishes correlations between the two, and the amount of information gained by the apparatus is closely related to the amount of correlation.

In quantum computation, we can think of one register holding the solution to our problem and another register querying it in order to extract the solution. The speed with which entanglement is established between the two determines the speed with which the problem is solved.

A physical system which has a finite amount of energy E and occupies a finite volume of space R^3 can hold only a finite amount of information and can process only it at a finite speed. The limits are given by the two bounds due to Bekenstein:

$$S \leq \frac{ER}{\hbar c}, \tag{12.17}$$

$$\frac{dS}{dt} \leq \frac{E}{2\hbar}. \tag{12.18}$$

Let us, just for fun, compute the amount of information that can potentially be stored in our heads. We can assume that the energy in a head is given by $E = mc^2$, where $m = 5$ kg on average. The average radius of a human head is about 0.1 m. The amount of information is therefore $S_{\text{head}} \approx 10^{42}$ bits of information, which is huge. A typical computer can at present store at most something like 10^{12} bits.

13
Quantum error correction

Quantum error correction is a very advanced field, and researchers have invested a great deal of time in trying to find the best ways of combating errors in quantum computers, which occur when the computer is not properly isolated from its environment. It is possible to show—theoretically speaking—that quantum computation can in principle be stabilized against any kind of influence of the environment. We shall learn about the basic techniques of quantum error correction in this chapter. It is, however, very much worth bearing in mind that a quantum computer may be implemented one day in a medium that is intrinsically fault-tolerant—just as the conventional classical computer is. The early pioneers of classical computers, such as von Neumann worked on classical error correction, which has been generalized to quantum mechanics to show that a reliable quantum computer can be constructed out of unreliable simple elements (gates).[1]

13.1 Introduction

In the previous chapters we have seen how to quantify entanglement, and how to use entanglement to compute. Now we turn our attention to realistic situations involving manipulations of quantum information, namely manipulations in noisy environments. In most realistic cases, quantum information is gradually lost owing to detrimental interactions with the environment. In this chapter we focus on methods of protection for quantum states in dissipative and decoherent environments.[2]

13.2 A simple example

A single qubit can be thought of as suffering three different types of errors, each represented by one Pauli operator. It is at first tempting to think that we cannot have quantum error correction as cloning of quantum states is not possible. Cloning of classical states is the key to classical error correction. The way error correction works classically is that instead of using one bit to encode zero or one, we use, for example, three bits. Then, if there is an error, we just need to look at all three bits and see what the majority of them encode. If two are in the state 1 (or 0), then (assuming that the

---

[1] Another important use was part of von Neumann's original intention. This was to show that unreliable machines can reproduce reliably, i.e. that life is possible according to the laws of (classical) physics!

[2] Dissipation implies loss of energy to the environment, while decoherence implies the loss of coherence, i.e. superpositions, and may not involve any energy exchange.

error probability is reasonably small) we can guess that the initial encoded state was 1 (or 0, respectively).

Redundancy in the information is the key to error correction (this is true in nature as well, for example in the encoding of amino acids by DNA. Here a number of different triplet bases may be associated with the same amino acid, so that even if there are errors in the replication of the DNA, this may not change the final synthesis of a protein.) Contrary to this logic, namely in spite of the "no cloning of quantum states", it turns out that we can perform quantum error correction; we do not need to clone states, but only on bits in each superposition element, and this is allowed by the quantum rules.

It is easy to protect against a single flip, represented by $\sigma_1 = |0\rangle\langle 1| + \langle 1||0\rangle$. We need to encode a superposition state of one qubit with two more qubits in the following way (the full network is given in Fig. 13.1):

$$\alpha|0\rangle + \beta|1\rangle \rightarrow \alpha|000\rangle + \beta|111\rangle. \tag{13.1}$$

After an error has occurred, we then need to apply a "coherent majority vote". This is performed in the following way. We first apply a controlled-not (CNOT) gate from the first to the second qubit and from the first to the third qubit. Then, we apply a Toffoli gate from the last two qubits to the first. The error will then be fully corrected. Let us see why this method works. If there are no errors, this procedure does not change the state of the first qubit. The same is true if errors occur in the second and third qubit. However, if there is a bit flip in the first qubit (the one that was originally encoded), then this is corrected by the application of the last Toffoli gate (readers are invited to confirm this for themselves).

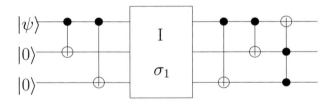

Fig. 13.1 A quantum network implementing an amplitude error correcting code. The initial state $|\psi\rangle$ is encoded into three bits using two CNOT gates (a CNOT gate is a controlled-NOT gate which applies NOT when the control bit is in the state $|1\rangle$). If at most one bit flip error occurs, indicated by the box, then the subsequent decoding and error correction using a Toffoli gate (which takes $|xyz\rangle$ into $|xy(z + x.y \bmod 2)\rangle$) restores the state $|\psi\rangle$ of the first qubit completely.

Protecting against a phase error (i.e. the action of the operator $\sigma_3 = |0\rangle\langle 0| - |1\rangle\langle 1|$) seems more complicated, but it is in fact at the same level of difficulty.[3] To protect against phase errors, we need to encode again with two extra qubits, but this time in the $|\pm\rangle$ basis:

---

[3] This was the biggest difficulty in constructing quantum error correction. The first codes were designed independently by Andrew Steane and Peter Shor.

$$\alpha|0\rangle + \beta|1\rangle \to \alpha|---\rangle + \beta|+++\rangle \tag{13.2}$$

The action of σ_3 on this rotated basis is the same as the action of σ_1 on the original basis (and we have seen how to deal with this). So the phase error is corrected by a coherent majority vote in the rotated basis.

The remaining error, $\sigma_2 = i|1\rangle\langle 0| - i|0\rangle\langle 1|$, is the product of the above two errors and can therefore be corrected simply by correcting for σ_3 and σ_1 in succession. Using an efficient implementation, the total number of qubits we need to protect against a single general error is $2 + 2 = 4$ qubits in addition to the original information-carrying qubit. If we have five qubits in total, then there are 3×5 single-qubit errors, since each qubit can suffer one of the above three errors. And there are $2^4 = 16$ orthogonal states to be occupied after errors. So all errors are distinguishable and can therefore be corrected. This idea for correcting single errors can be generalized to any number of errors and we can, in a similar vein, derive a formula for the minimal number of qubits needed.

13.3 General conditions

In reality, the main source of coherence loss is dissipative coupling to an environment with a large number of degrees of freedom, which must be traced out of the problem. This loss is often manifested as some form of spontaneous decay, whereby quanta are randomly lost from the system. Each interaction with, and hence each instance of dissipation to, the environment can be viewed in information-theoretic terms as introducing an error in the measurement of the output state. There are, however, techniques for 'correcting' errors in quantum states. The basic idea of error correction is to introduce an excess of information, which can then be used to recover the original state after an error. These quantum error correction procedures are themselves quantum computations, and as such are also susceptible to the same errors. This imposes limits on the nature of the "correction codes", which are explored in this section.

First we derive some general conditions which a quantum error correction code has to satisfy.[4] Assume that q qubits are encoded in terms of $n \geq q$ qubits to protect against a certain number of errors d. We construct 2^q codewords, each being a superposition of states that have n qubits. These code words must satisfy certain conditions, which are derived in this section. There are three basic errors (i.e. all other errors can be written as a combination of these): an amplitude error, $\hat{A}$, which acts as a NOT gate; a phase error, $\hat{P}$, which introduces a minus sign into the upper state; and the combination of these, $\hat{A}\hat{P}$. A subscript will be used to designate the position of the error, so that $\hat{P}_{1001}$ means that the first and the fourth qubit undergo a phase error. These errors are the same as the combinations of Pauli operators considered the previous section.

We consider an error to arise because of the interaction of the system with a "reservoir" (i.e. any other quantum system), which then becomes entangled with it. This procedure is the most general way of representing errors, which are not restricted to discontinuous "jump" processes, but encompass the most general type of interaction.[5]

---

[4]These conditions were first derived by Knill and Lafflamme, although here I follow a completely different approach.

[5]There are actually no jumps in nature, as far as we understand.

Error correction is thus seen as a process of disentangling the system from its environment back to its original state. The operators $\hat{A}$ and $\hat{P}$ are constructed so as to operate only on the system, and are defined in the same way as the operators for a complete measurement. In reality, each qubit would couple independently to its own environment (though not assuming this would not change any part of the analysis significantly), so the error in a given state could be written as the direct product of the errors in the individual qubits. A convenient error basis for a single error on a single qubit is $\{\hat{I}, \hat{\sigma}_i\}$, where the $\hat{\sigma}_i$'s are the Pauli matrices. In this case, the error operators are Hermitian, and square to the identity operator, and we assume this property for convenience throughout the following analysis.

In general, the initial state can be expressed as

$$|\psi_i\rangle = \sum_{k=1}^{2q} c_k |C^k\rangle |R\rangle , \qquad (13.3)$$

where the $|C^k\rangle$ are the codewords for the states $|k\rangle$, and $|R\rangle$ is the initial state of the environment.[6] The state after a general error is then a superposition of all possible errors acting on the above initial state,

$$|\psi_f\rangle = \sum_{\alpha\beta} \hat{A}_\alpha \hat{P}_\beta \sum_k c_k |C^k\rangle |R_{\alpha,\beta}\rangle , \qquad (13.4)$$

where $|R_{\alpha,\beta}\rangle$ is the state of the environment.[7] Note that $|R_{\alpha,\beta}\rangle$ depends only on the nature of the errors, and is **independent** of the codewords. The above is, in general, not in the Schmidt form, that is, the codeword states after the error are not necessarily orthogonal and neither are the states of the environment.

Now, since we have no information about the environment, we must trace it out using an orthogonal basis for the environment $\{|R_n\rangle, n = 1, d\}$. The resulting state is a mixture of the form $\hat{\eta}_i = \sum_n |\psi_n\rangle\langle\psi_n|$, where

$$|\psi_n\rangle = \sum_{\alpha\beta} x_n^{\alpha\beta} \hat{A}_\alpha \hat{P}_\beta \sum_k c_k |C^k\rangle , \qquad (13.5)$$

and $x_n^{\alpha\beta} = \langle R_n | R_{\alpha\beta}\rangle$. To detect an error, we then perform a measurement on the state $\hat{\eta}$ to determine whether it has an overlap with one of the following subspaces:

$$H_{\alpha\beta} = \{\hat{A}_\alpha \hat{P}_\beta |C^k\rangle, k = 1, \ldots, 2^q\} . \qquad (13.6)$$

The initial space, after the error, is given by the direct sum of all the above subspaces, $H = \sum_{\alpha\beta} \oplus H_{\alpha\beta}$. Each time we measure an overlap and obtain a zero result, the state space H reduces in dimension, eliminating that subspace as containing the state after the error. Eventually, one of these overlap measurements will give a positive result, which is mathematically equivalent to projecting onto the corresponding subspace. The

---

[6]We consider the most general possible superpositions of code words.

[7]We keep the hat notation to designate operators in this section in order to avoid any confusion.

state after this projection is then given by the mixture $\hat{\eta}_f = \sum_n |\psi_{n\text{Proj}_{\alpha\beta}}\rangle\langle\psi_{n\text{Proj}_{\alpha\beta}}|$, where

$$|\psi_{n\text{Proj}_{\alpha\beta}}\rangle = \sum_{kl}\sum_{\gamma\delta} x_n^{\gamma\delta} \hat{A}_\alpha \hat{P}_\beta |C^k\rangle\langle C^k|\hat{P}_\beta \hat{A}_\alpha \hat{A}_\gamma \hat{P}_\delta |C^l\rangle c_l \ . \qquad (13.7)$$

A successful projection will effectively take us to a state generated by a superposition of certain types of error. One might expect that to distinguish between various errors, the different subspaces $H_{\alpha\beta}$ would have to be orthogonal. However, we shall show that this is not, in fact, necessary (although it is obviously sufficient).

After having projected onto the subspace $H_{\alpha\beta}$, we now have to correct the corresponding error by applying the operator $\hat{P}_\beta \hat{A}_\alpha$ to $|\psi_{\text{Proj}_{\alpha\beta}}\rangle$, since $\hat{P}_\beta \hat{A}_\alpha \hat{A}_\alpha \hat{P}_\beta = \hat{1}$. In order to for us correct the error successfully, the resulting state has to be proportional to the initial state of the codewords in $|\psi_i\rangle$. This leads to the condition

$$\sum_{kl}\sum_{\gamma\delta} x_n^{\gamma\delta} |C^k\rangle\langle C^k|\hat{P}_\beta \hat{A}_\alpha \hat{A}_\gamma \hat{P}_\delta |C^l\rangle c_l = z^{\alpha\beta n}\sum_m c_m |C^m\rangle \ , \qquad (13.8)$$

where $z^{\alpha\beta n}$ is an arbitrary complex number. Now we use the fact that all code words are mutually orthogonal, that is, $\langle C^k|C^l\rangle = \delta_{kl}$, to obtain the result that

$$\sum_l \sum_{\gamma\delta} c_l x_n^{\gamma\delta} \langle C^k|\hat{P}_\beta \hat{A}_\alpha \hat{A}_\gamma \hat{P}_\delta |C^l\rangle = z^{\alpha\beta n} c_k \qquad (13.9)$$

for all k and arbitrary c_k. This can be written in matrix form as

$$\mathbf{F}^{\alpha\beta n}\mathbf{c} = z^{\alpha\beta n}\mathbf{c} \ , \qquad (13.10)$$

where the elements of the matrix $\mathbf{F}$ are given by

$$F_{kl}^{\alpha\beta n} := \sum_{\gamma\delta} x_n^{\gamma\delta} \langle C^k|\hat{P}_\beta \hat{A}_\alpha \hat{A}_\gamma \hat{P}_\delta |C^l\rangle \ . \qquad (13.11)$$

As eqn 13.10 is valid for all $\mathbf{c}$, it follows that

$$\forall \ k,l, \quad F_{kl}^{\alpha\beta n} = z^{\alpha\beta n}\delta_{kl} \ . \qquad (13.12)$$

However, we do not know the form of the $x_n^{\gamma\delta}$'s, as we have no information about the state of the environment. Therefore, for the above to be satisfied for *any* form of x's, we need each individual term in eqn 13.11 to satisfy

$$\langle C^k|\hat{P}_\beta \hat{A}_\alpha \hat{A}_\gamma \hat{P}_\delta |C^l\rangle = y^{\alpha\beta\gamma\delta}\delta_{kl} \qquad (13.13)$$

where $y^{\alpha\beta\gamma\delta}$ is *any* complex number. From eqns 13.11–13.13, we see that the numbers x, y, and z are related through

$$\sum_{\gamma\delta} x_n^{\gamma\delta} y^{\alpha\beta\gamma\delta} = z^{\alpha\beta n} \ . \qquad (13.14)$$

Equation 13.13 is the main result in this section, and gives a general, and in fact the *only*, constraint on the construction of codewords, which may then be used for encoding purposes. If we wish to correct for up to d errors, we have to impose a further constraint on the subscripts α, β, γ, and δ; namely, wt(supp(α) $\cup$ supp(β)), wt(supp(γ) $\cup$ supp(δ)) $\leq d$, where supp(x) denotes the set of locations where the n-tuple x is different from zero and wt(x) is the Hamming weight, that is, the number of digits in x different from zero. This constraint on the indices of the errors simply ensures that they do not contain more than d logical 1's altogether, which is, in fact, equivalent to no more than d errors occurring during the process.

We emphasize that these conditions are the most general possible. By substituting $z^{\alpha\beta\gamma\delta} = \delta_{\alpha\beta}\delta_{\gamma\delta}$ in eqn 13.13, we obtain the conditions

$$\langle C^k|\hat{P}_\beta \hat{A}_\alpha \hat{A}_\gamma \hat{P}_\delta|C^l\rangle = \delta_{\beta\delta}\delta_{\alpha\gamma}\delta_{kl} \ . \tag{13.15}$$

This special criterion says that the states $\hat{A}_\gamma \hat{P}_\delta|C^l\rangle$ and $\hat{A}_\alpha \hat{P}_\beta|C^k\rangle$ have to be orthogonal to each other unless $\alpha = \gamma$, $\beta = \delta$, and $k = l$. Thus, if $k \neq l$ (i.e. the code words are different), then the errors should lead to orthogonal states. And this is the same as in classical error correction. The general conditions, therefore, show the main difference between quantum and classical error correction: it is possible for two different errors to lead to the same state providing that the overlap is the same for all the codewords. Classically, on the other hand, different errors must lead to completely different states in order for us to have successful error correction.

13.4 Reliable quantum computation

We have seen how to protect qubits against general errors. However, this protection is rather "static", that is, our qubits are not evolving while errors occur. Suppose we wish to implement a controlled-NOT between two qubits which can undergo an error during this operation. Is there a point to encoding these qubits in the first place, since the encoding and decoding procedures are just composed of a number of CNOT (and other) gates, which themselves can undergo errors? It appears that if (realistically) we allow encoding and decoding to undergo errors, then there is no point in protecting gates, since this action introduces even more errors. The conclusion would be that quantum error correction cannot be used in quantum computation!

The same conclusion was reached in the 1930s about classical computation. Then, however, von Neumann showed this to be a completely erroneous conclusion, and he proved that reliable computation (classical of course, as von Neumann did not know about quantum computation) is possible from unreliable components. His argument can be translated directly into quantum computing and this gives rise to the fault-tolerant quantum computation, that is, in von Neumann's jargon, reliable quantum computation with unreliable components. We now present a sketch of this argument. This is intended only as a qualitative argument that the quantum error correction we have studied in this chapter can be applied to quantum computing in general, and no details will be given.

The idea of fault-tolerant quantum computation is to encode the qubits in such a way that the encoding does not introduce more errors than were previously present. If the error stays at the same level, we keep performing error correction until the error has decreased in magnitude. We have seen that (statically) it requires five qubits to encode a single qubit against a single error. It is the iterative application "in depth" of the encoding that will enable us to reduce errors to an arbitrarily small level, providing they are below a certain level to start with. In other words, we shall be encoding the encoding bits. Before we give more details, let us just recapitulate the main points about a quantum computer.

The input to a quantum computer is a string of qubits. For the following calculation, a quantum computer is viewed as consisting of two main parts: *quantum gates* and *quantum wires*. By *basic* quantum gates, we mean any set of quantum gates which can perform any desired quantum computation. A universal quantum gate is a gate such that a combination of them can be used to simulate any other quantum gate. A quantum wire is used as a representation of that part of the computation of any qubit where the evolution is a simple identity operation (i.e. no gate operates on the qubit), as well as of the time the qubit spends during the operation of a gate.

For stable quantum computation, we obviously require that the probability of error after a fault-tolerantly encoded basic gate is of higher order (i.e. the error is smaller) than the probability of error after an unencoded gate (that is the whole point of encoding and fault-tolerant error correction!). From this, we can derive a bound on the size of the allowed errors in the wires and gates. When we encode the encoding bits again, we reduce the error further and can reduce the error arbitrarily for an arbitrarily long computation. Therefore, given certain initial limits on the error rate in the gates and wires we can stabilize any computation to an error rate as small as we desire, given an unlimited amount of time. Consider a two-input, two-output quantum gate. The probability of having any of the three basic errors in the first or the second wire is η—giving the overall first-order wire probability of error as 2η. The error in the gate itself is ϵ. We assume that the overall probability of error for the whole basic gate is less than or equal to $2\eta + \epsilon$. Suppose that the basic gate is now encoded fault-tolerantly against a single error of any kind, using l qubits. The overall second-order error at the end of the gate is then

$$\eta^*(\eta, \epsilon, l) = \left(1 - \frac{l(l-1)}{2} l^4 \eta^2\right) l^2 \epsilon + \frac{l(l-1)}{2} l^4 \eta^2 (1 - l^2 \epsilon) , \qquad (13.16)$$

that is, it is equal to the probability of having an error in the wires (this time to second order) and not in the gates, plus that of having error an in the gates and not in the wires. The term $l(l-1)/2$ comes from choosing two out of l gates where an error can occur, and the factor l^4 derives from the use of l^2 gates, so that the error is transformed according to $\eta \to l^2 \eta$ and is of second order.

We require that fault-tolerant error correction reduces the error. Hence

$$\left(1 - \frac{l(l-1)}{2} l^4 \eta^2\right) l^2 \epsilon + \frac{l(l-1)}{2} l^4 \eta^2 (1 - \epsilon) \leq 2\eta + \epsilon , \qquad (13.17)$$

As the LHS is $> \eta$, we simplify the above without a greater loss in generality to

$$\left(1 - \frac{l(l-1)}{2} l^4 \eta^2\right) l^2 \epsilon + \frac{l(l-1)}{2} l^4 \eta^2 (1 - \epsilon) \leq \eta . \qquad (13.18)$$

The solutions to the equation derived from the above are

$$\eta_{\pm} = \frac{1 \pm \sqrt{1 - 2(l^8\epsilon - 2l^{10}\epsilon^2)}}{(l^6 - 2l^8\epsilon)}. \qquad (13.19)$$

We require that $\eta \in R$ (and that $0 \leq \eta \leq 1/2$), so that we have the following two regimes of error:

1. $0 < \eta < \eta_+$ and $e \leq e_-$.
2. $0 < \eta < \eta_-$ and $e \geq e_+$.

Here, $\epsilon_\pm = 1/(2l^2)(1 \pm \sqrt{1 - 2l^{-6}})$. The output of the first encoded basic gate is fed into the next one (or part of the output is fed into one next basic gate and the rest into another basic gate). It is evident that if condition 1 holds, further encoding can only decrease the error. The residual error not taken into account is $\sim l^3(l^2\eta)^3 = l^9\eta^3$ (i.e. the second-order error is not corrected by our encoding). In the worst case, when $\epsilon = \epsilon_- \sim l^{-8}$, we get $\eta \sim l^{-6}$, which means that the residual uncorrected error is $\sim l^{-9}$. This error can accumulate over time if the computation is sufficiently long. However, the residual error after n in-depth encodings is $l^{-O(n)}$, which can be made arbitrarily small using a sufficiently large n. Therefore if the initial error per gate is sufficiently small, the gates can be used to perform arbitrary large quantum computations. If we need $l = 10$ qubits to fault-tolerantly encode one qubit, then the tolerable error rate is 10^{-6} which a more careful analysis shows to be correct. By using some further tricks, we can increase this tolerable rate to a probability of error per elementary gate 10^{-3}. At present we are still far away from this requirement. Typically we would have an error rate of 90–95% of error per gate even in the most reliable implementations, which is 50 times more than required for fault tolerance.

13.5 Quantum error correction considered as a Maxwell's demon

The idea that quantum error correction reverses the effects of noise from the environment sounds like a kind of Maxwell's demon, able to counter the environmentally driven entropy increase in a system. In this chapter, we in fact show that this analogy can be made perfectly precise.[8]

In order to link information theory to thermodynamics we shall use a box containing a single atom as a representation of a classical bit of information:[9] if the atom is in the left-hand half (LHH), it will represent a 0, and if it is in the right-hand half (RHH), it will represent a 1. Now, if the atom is already confined to one of these halves, and we expand it isothermally and reversibly to occupy the whole volume, then the entropy of the atom increases by $\Delta S = k \log 2$ and the free energy decreases by $\Delta F = -kT \log 2$. The atom does an amount of work $\Delta W = kT \log 2$. Suppose that, initially, we want to have our atom in one of the halves in order to be able to do some work as described. However, suppose also that there is a possibility of error, namely

---

[8] For a collection of important papers on the subject of Maxwell's demon see, Leff and Rex, (2003)

[9] Here we use a simple reversible cycle, which is a slight modification of Bennett's (1982) version of Maxwell's demon, to illustrate the process of error correction. Bennett's original motivation was to show that Maxwell's demon cannot beat the Second Law, simply because anything that the demon memorizes during the cycle has to be erased in the end. Our motivation here is different, but error correction of course, also has to balance the books as far as the Second Law is concerned.

168 *Quantum error correction*

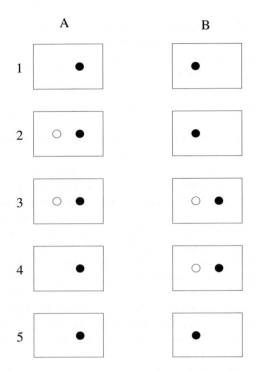

Fig. 13.2 Classical error correction considered as a Maxwell's demon. The steps are detailed in the text and their significance is explained. (1) The states of atoms A and B are initially uncorrelated; (2) atom A undergoes an error; (3) atom B observes atom A, and the atoms thereby become correlated; (4) atom A is corrected to its initial state and the atoms are now uncorrelated; (5) atom B is returned to its initial state and the whole cycle can start again.

the atom has a chance of $1/2$ of jumping to the other half. Once this happens we cannot extract any work until we return the atom to the initial state. But this itself requires an amount of work $kT \log 2$ in an isothermal compression. We would thus like to be able to correct this error, and so we introduce another atom in a box to monitor the first one. This is represented in Fig. 13.2, and the whole error correction protocol goes through five stages:

1. Initially the atoms are in LHH and RHH of respective boxes.
2. An error then affects atom A so that it now has a 50/50 chance of being in the LHH or the RHH.
3. Atom B observes atom A and becomes correlated with it, so that either both occupy the LHHs or both occupy the RHHs. We make no assumptions about how the observation is made.
4. Depending on the state of B, we now compress A into one of the two halves; this involves no work, but the state of B is now not known—it has a 50/50 chance of being in either the LHH or the RHH. Thus we have corrected A at the cost of randomizing B. It should be pointed out that by "work", we always mean here

the work done by the atom (or on the atom) on the piston (or by the piston). As is usual in thermodynamic idealizations of this kind, all other work is neglected (or assumed to be negligible). For example, the partition itself is assumed to be very light (in fact, to have zero mass), so that no work is done in pushing it. Here, no work done by the atom, since it is not contained in that part of the box which is compressed (this information about the position of A is recorded by B).

5. In order to be able to repeat the error correction, we need to reset B to its initial state as in step 1. Thus we perform an isothermal, reversible compression on B.

Let us now analyze this process using entropy and free energy. In step 1, both of the atoms possess $kT \log 2$ of free energy. After A undergoes an error, its free energy is decreased by $\Delta F_A = -kT \log 2$, and nothing happens to B. The total free energy is now $\Delta F_{AB} = kT \log 2$. In step 3, the total free energy is still $\Delta F_{AB} = kT \log 2$, but the atoms are correlated. This means that atom B has information about A (and vice versa). The amount of information is $k \log 2$. This enables the error correction step to take place in step 4. This does not change the total free energy, but the atoms are now decorrelated. In step 5, a quantity of work $kT \log 2$ is invested in resetting the state of atom B so that the initial state in step 1 is reached. This completes the cycle, which can now start again. What happens to the entropy? The entropy of each atom is initially 0. Then an error increases the entropy of A to $\Delta S_A = k \log 2$. In step 3, the atoms become correlated so that they both have the same entropy, that is, $\Delta S_B = k \log 2$. However, the crucial point is that the total entropy does not change from step 2 to step 3. This is the point of observation, and the information gained by B about the state of A is $S_A + S_B - S_{AB} = k \log 2$. In step 4, $\Delta S_A = -k \log 2$ and there are no changes for atom B. In the resetting step, $\Delta S_B = -k \log 2$, so that now both of the atoms have an entropy of zero as at the beginning. Another change that has taken place, and this is the crux of Landauer's principle, is that in the compression of atom B, work was invested and the entropy of the environment increased by $k \log 2$. This final entropy increase is necessary for resetting and is in this case equal to the amount of information gained in step 3. Landauer's principle of erasure states that the entropy wasted in resetting is at least as big as the information gain. If this were not so, we could use the above cycle to do work by extracting heat from the environment with no other changes and the Second Law of Thermodynamics (in Kelvin's form stating that not all heat can be converted fully into work) would be violated. Thus, here an error means that atom A's ability to do work has been destroyed and in order to correct this we need another atom B to transfer its free energy to A. In this process atom B loses its ability to do work and, in order to regain it, an amount of $k \log 2$ of entropy has to be wasted (thus "saving" the Second Law).

We now turn to quantum error correction. The aim of quantum error correction as presented in this section will be to preserve a given quantum state of a quantum mechanical system, much as a refrigerator is meant to preserve the low temperature of food in a higher-temperature environment (a room). Some work is performed in the refrigerator which then reduces the entropy of the food by increasing the entropy of the surroundings. In accordance with the Second Law, the entropy increase of the environment is at least as large as the entropy decrease of the food. Analogously, when there is an error in the state of a quantum system, the entropy usually increases (this is, however, not always the case, as we shall see later), and the error correction reduces it back to the original state thereby decreasing the entropy of the system, but increasing the entropy of the environment (or what we shall call a "garbage can"). To quantify this precisely, let us look at the error correction process in detail.

Suppose we wish to protect a pure state $|\psi\rangle = \sum_i c_i |a_i\rangle$, where the $\{|a_i\rangle\}$ form an orthonormal basis. This is usually done by introducing redundancy, that is, by encoding the state of a system in a larger Hilbert space according to some rule

$$|a_i\rangle \rightarrow |C_i\rangle , \qquad (13.20)$$

where the $\{|C_i\rangle\}$ are the code words. Note that this step was omitted in the classical case. This is because the very existence of system B can be interpreted as encoding. The main difference between classical and quantum error correction is that errors in the classical case can always be distinguished. In quantum mechanics these can lead to nonorthogonal states, so that errors cannot always be distinguished and corrected. So it might be said that the encoding in quantum mechanics makes errors orthogonal and hence distinguishable (a precise mathematical statement of this was given earlier). Of course, redundancy also exists in classical error correction (above, we have another system, B, to protect A), but the states are already orthogonal and distinguishable by the very nature of being classical. In the quantum case we introduce an additional system, called the apparatus, in order to detect different errors; this plays the role that B plays in classical error correction. Now the error correction process can be viewed as a series of steps. First, the initial state is

$$|\psi_c\rangle |m\rangle |e\rangle , \qquad (13.21)$$

where $|\psi_c\rangle = \sum_i c_i |C_i\rangle$ is the encoded state, $|m\rangle$ is the initial state of the measuring apparatus, and $|e\rangle$ is the initial state of the environment. Now, the second stage is the occurrence of errors, represented by operators $\{E_i\}$ which act on the state of the system only, after which we have

$$\sum_i E_i |\psi_c\rangle |m\rangle |e_i\rangle . \qquad (13.22)$$

Note that at this stage the measurement has not yet been made so that the state of the apparatus is still disentangled from that of the rest. In general, the states of the environment $|e_i\rangle$ need not be orthogonal (as we have seen). If they are orthogonal, this leads to a specific form of decoherence, which we might call "dephasing" and will be analyzed later. However, the formalism we present here is completely general and applies to any form of error. Now the measurement occurs, and we obtain

$$\sum_i E_i |\psi_c\rangle |m_i\rangle |e_i\rangle . \qquad (13.23)$$

The error correction can be seen as an application of E_i^{-1}, conditional on the state m_i. (This, of course, cannot always be performed, but the code words have to satisfy the conditions earlier in order to be correctable. Here we need not worry about this;

our aim is only to understand the global features of error correction). After this, the state becomes

$$|\psi_c\rangle \sum_i |m_i\rangle |e_i\rangle , \quad (13.24)$$

and the state of the system returns to the initial encoded state; the error correction has worked. However, notice that the state of the apparatus and the environment is not equal to the initial state. This feature will be dealt with shortly. Before that, let us note that the total state (system + apparatus + environment) is always pure. Consequently, the von Neumann entropy is always zero. Therefore it is difficult to see how this process can be compared with refrigeration, where the entropy is kept low at the cost of an increase in the entropy of the environment. However, in general, the environment is not accessible and we usually have no information about it (if we had this information we would not need error correction!). Thus the relevant entropies are those of the system and the apparatus. This means that we can *trace out* the state of the environment in the above picture; this leads to dealing with mixed states and increasing and decreasing entropies. In addition, the initial state of the system might be pure or mixed (above, we have assumed a pure state), and we now analyze these two cases separately.

13.5.1 Pure states

We follow the above set of steps, but now the environment will be traced out of the picture after errors have occurred. Therefore in the first step the state is as follows:

1. After the errors, the state is $\sum_i E_i |\psi_c\rangle |m\rangle |e_i\rangle$, where we have assumed the "perfect" decoherence, that is, $\langle e_j | e_i \rangle = \delta_{ij}$, for simplicity.

2. Now the environment is traced out, leading to $\sum_i E_i |\psi_c\rangle \langle\psi_c| E_i^\dagger \otimes |m\rangle \langle m|$. Note that this is not a physical process, just a mathematical way of neglecting a part of the total state (we have introduced the direct-product sign just to indicate the separation between the system and the apparatus; when there is no possibility of confusion, we shall omit it).

3. Then the system is observed, thus creating correlations between the apparatus and the system, leading to $\sum_i E_i |\psi_c\rangle \langle\psi_c| E_i^\dagger \otimes |m_i\rangle \langle m_i|$. We assume that the observation is perfect so that $\langle m_j | m_i \rangle = \delta_{ij}$. Note that we need different errors to lead to orthogonal states if we wish to be able to correct them. Here, also, if the observation is imperfect then error correction cannot be completely successful, since nonorthogonal states cannot be distinguished with perfect efficiency.

4. The correction step happens and the system is decorrelated from the apparatus so that we have $|\psi_c\rangle \langle\psi_c| \otimes \sum_i |m_i\rangle \langle m_i|$. As we have remarked before, this is not equal to the initial state of the system and the apparatus. If we imagine that we have to perform correction a number of times in succession, then this state of the apparatus would not be helpful at all. We need to somehow reset it back to the original state $|m\rangle$.

5. This is done by introducing another system, called a garbage can (gc), which is in the correct state $|m\rangle$, so that the total state is $|\psi_c\rangle \langle\psi_c| \otimes \sum_i |m_i\rangle \langle m_i| \otimes |m\rangle \langle m|$, and then swapping the states of the garbage can and the apparatus (this can be

performed unitarily) so that we finally obtain $|\psi_c\rangle\langle\psi_c| \otimes |m\rangle\langle m| \otimes \sum_i |m_i\rangle\langle m_i|$. Only now are the system and the apparatus ready to undergo another cycle of error correction.

We can now apply an entropy analysis to this error correction cycle. In the first step, the entropy of the system + apparatus increases by $\Delta S_{S+A} = S(\varrho)$, where $\varrho = \sum_i E_i |\psi_c\rangle\langle\psi_c| E_i^\dagger$. Step 2 is not a physical operation, and so there is no change in entropy. In the third step, there is also no change in entropy; the only change is the correlations between the system and the apparatus are created. Step 4 is similar to step 2, and so there is no change in entropy overall. In step 5, the entropy of the system + apparatus is zero since they are in a pure total state. Thus, $\Delta S_{S+A} = -S(\varrho)$, and now we see the formal analogy with the refrigeration process: the net change in entropy of the system + apparatus is zero, and the next error correction step can begin; however, at the end the gc has increased its entropy by $\Delta S_{gc} = S(\varrho)$. This is now exactly a manifestation of Landauer's erasure principle. The information gain in step 3 is equal to the mutual entropy between the system and the apparatus, $I_{S+A} = S_S + S_A - S_{S+A} = S(\varrho)$. The logic behind this formula is that before the observation the apparatus did not know anything about the system, and therefore the state of the system's state was uncertain by $S(\varrho)$, whereas after the observations this uncertainty is zero—the apparatus knows everything about the system. We note that this information is the Shannon mutual information between the two (Schmidt) observables pertaining to the system and the apparatus. This needs to be erased at the end to start a new cycle and the entropy increase is exactly (in this case) equal to the information gained. So from the entropic point of view, we have performed the error correction in the most efficient way, since, in general, the entropy of the gc increase is larger than the information gained.

Next we consider the correction of mixed states. This might at first appear useless, because we might think that a mixed state is one that has already undergone an error. This, however, is not necessarily so, and this situation occurs, for example when we are protecting a part of an entangled bipartite system. It might be thought that an analogous case does not exist in the case of classical error correction. There, an error was represented by a free expansion of atom A from step 1 to step 2. However, we could have equally well started from a situation where A occupied the whole volume, and treated an error as a "spontaneous" compression of the atom into one of the halves. If A was correlated with some other atom C (so that they both occupied the LHH or both occupied the RHH), then this compression would result in decorrelation, which is really an error. Thus classical and quantum error correction are in fact very closely related, which is also shown by their formal analogy to Maxwell's demon.

13.5.2 Mixed states

Now suppose that systems A and B are entangled and that we are only performing error correction on system A. In the case we are considering here, this is the same as protecting a mixed state. We are not saying that protecting a mixed state is in general the same as protecting entanglement. For example, if a state $|00\rangle + |11\rangle$ flips to $|01\rangle + |10\rangle$ with probability $1/2$, then the entanglement is destroyed, but the reduced states of each subsystem are still preserved. What we mean is that quantum error correction has been developed here to protect any pure state of a given system. In that case, any mixed state is also protected, and also any entanglement that it might have with other systems. This also means that, using standard quantum error correction, an entangled pair can be preserved just by protecting each of the subsystems separately. Now, for simplicity, let us say that we have a mixture of two orthogonal states $|\psi\rangle, |\phi\rangle$. The initial state is then, without normalization (and without the system B),

$$(|\psi\rangle\langle\psi| + |\phi\rangle\langle\phi|) \otimes |e\rangle\langle e| \otimes |m\rangle\langle m| . \qquad (13.25)$$

Now we can go through all the above stages:

1. Error: $\sum_i E_i(|\psi\rangle\langle\psi| + |\phi\rangle\langle\phi|)E_i^\dagger \otimes |e_i\rangle\langle e_i| \otimes |m\rangle\langle m|$;.
2. Tracing out the environment: $\sum_i E_i(|\psi\rangle\langle\psi| + |\phi\rangle\langle\phi|)E_i^\dagger \otimes |m\rangle\langle m|$;.
3. Observation: $\sum_i E_i(|\psi\rangle\langle\psi| + |\phi\rangle\langle\phi|)E_i^\dagger \otimes |m_i\rangle\langle m_i|$;.
4. Correction: $(|\psi\rangle\langle\psi| + |\phi\rangle\langle\phi|) \otimes \sum_i |m_i\rangle\langle m_i|$;.
5. Resetting: $(|\psi\rangle\langle\psi| + |\phi\rangle\langle\phi|) \otimes \sum_i |m_i\rangle\langle m_i| \otimes |m\rangle\langle m|$
  $\to (|\psi\rangle\langle\psi| + |\phi\rangle\langle\phi|) \otimes |m\rangle\langle m| \otimes \sum_i |m_i\rangle\langle m_i|$.

The entropy analysis is now as follows. In step 1, $\Delta S_{S+A} = S(\varrho_f) - S(\varrho_i)$, where $\varrho_f = \sum_i E_i(|\psi\rangle\langle\psi| + |\phi\rangle\langle\phi|)E_i^\dagger$ and $\varrho_i = |\psi\rangle\langle\psi| + |\phi\rangle\langle\phi|$ (not normalized). In steps 2 and 3, there is no change in entropy, although in step 3 an amount of information $I = S(\varrho_f)$ is gained if the correlation between the system and the apparatus is perfect (i.e. $\langle m_j|m_i\rangle = \delta_{ij}$). In step 4, $\Delta S_{S+A} = S(\varrho_i)$ as the system and the apparatus become decorrelated. In step 5, $\Delta S_{S+A} = -S(\varrho_f)$, but the entropy of the gc increases by $S(\varrho_f)$. Thus, altogether, $\Delta S_{S+A} = 0$, and the entropy of the gc has increased by exactly the same amount as the information gained in step 3, thus confirming Landauer's principle again.

Now we analyze what happens if the observation in step 3 is imperfect. Suppose for simplicity that we have only two errors, E_1 and E_2. In this case there are only two states of the apparatus, $|m_1\rangle$ and $|m_2\rangle$; an imperfect observation would imply that $\langle m_1|m_2\rangle = a > 0$. Now the entropy of the information erasure is $S(|m_1\rangle\langle m_1| + |m_2\rangle\langle m_2|)$, which is smaller than when $|m_1\rangle$ and $|m_2\rangle$ are orthogonal. This implies, via Landauer's principle, that the information gained in step 3 would be smaller than when the states of the apparatus are orthogonal, and this in turn leads to imperfect error correction. Thus, doing perfect error correction without perfect information gain is forbidden by the Second Law of Thermodynamics via Landauer's principle.[10]

13.6 Summary

In this chapter, I have tried to acquaint the reader with some of the most basic techniques of quantum error correction. We have derived general conditions for quantum errors and code words to be correctable.

The main conclusion was that it is possible to develop fault-tolerant quantum computation from faulty quantum gates. The threshold level of errors per gate was found to be 10^{-4}, which is four orders of magnitude lower than the best experiments at present (2006).

---

[10] This is analogous to von Neumann's (1952) proof that being able to distinguish perfectly between two nonorthogonal states would lead directly to violation of the Second Law of Thermodynamics.

Finally, error correction was found to be a type of Maxwell's demon. Here again we meet Landauer's erasure principle. Errors increase disorder (entropy) and in order to correct them—that is, to decrease the entropy of the system after it has undergone errors—we need to increase the entropy of the environment. Overall, the whole process is unitary and the entropy of the entire universe does not change.

14
Outlook

In this book I have tried, to the best of my ability, to cover the foundations of quantum information science but at the same time discuss the relationship between physics and information theory in general. This field is still relatively young and you will notice that many times I have raised questions which we are unable to answer at present. It is precisely this that makes it attractive to students, but this also makes it attractive to researchers in other fields. During the past ten years or so, I have seen mathematicians, computer scientists, high-energy physicists, quantum opticians and many others switch from their previous activities to investigating quantum information science. This is another reason why this field is so exciting: it is highly multidisciplinary and all these researchers with different backgrounds can learn a great deal from talking to each other.

To summarize the book, we now have a pretty good understanding of the quantum equivalents of the Shannon coding and channel capacity theorems, although there are still many open problems here related to the details of quantum information transfer. We have seen that the von Neumann entropy plays a role analogous to the Shannon entropy and that the Holevo bound is the analogue of Shannon's mutual information used to quantify the capacity of a classical channel. Furthermore, we have seen that quantum systems can process information more efficiently than classical systems in a number of different ways. We have learnt about quantum teleportation and quantum dense coding, both of which rely on entanglement.

Entanglement is an excess of correlations that can exist in quantum physics and is impossible to reproduce classically (with what we have called "separable" states). We have seen how to discriminate entangled from separable states using entanglement witnesses, as well as how to quantify entanglement. Here we have also encountered a number of exciting open problems, one very general one being the possibility and extent of the existence of macroscopic entanglement at high temperatures. And, even more excitingly, can entanglement be used by living organisms? This has brought us directly to the forefront of research activities in the field.

Entanglement is also believed to be responsible for the speedup in quantum computation over its classical counterpart. This statement, although very natural, has not yet been proved in a rigorous way (though there are many arguments in its favor). We have reviewed the two major quantum algorithms—a factorization and a search algorithm—and showed how quantum superpositions are exploited there to achieve a considerable speedup. Finally, we have reviewed how quantum computation can be executed in practice even when the fundamental building blocks of quantum computers—the basic gates and qubits—are not perfect. There we paralleled an old argument of von Neumann, which he used to show that fault-tolerant classical computation can be achieved using faulty building blocks. It is comforting to see that the same holds in quantum computation.

As I write this final chapter (in the summer of 2005), actual quantum computers have reached a size of about $10 - 15$ qubits. It seems to be very difficult at present to proceed experimentally beyond this number. A lot of ingenuity is, however, going into proposing new implementations and media for quantum computers (ranging from optical to solid-state), so that it seems to me to be only a matter of time before quantum computers appear on your desktop.

The practical implementation of quantum computation will be an exciting research direction for a long time to come. What about new theoretical directions that are likely to spring up in the future? If I were to speculate, I would bet on more fundamental results emerging from our understanding of quantum entanglement. Correlations are very fundamental for our description and understanding of the world. It is, in fact, tempting to say that things and events have no meaning in themselves, but that only correlations between them are "real". This philosophy is known under the general name of "relationalism". Einstein's theory of general relativity is a shining example of the philosophy of relationalism put into practice in physics. Einstein was, however, unable to take relationalism one step further, to find a unified framework for general relativity and quantum physics. To this day, this task remains an open problem—arguably the most important one in physics at present. Could it be that this unification is proving so elusive because quantum correlations (entanglement) have not been incorporated at the most fundamental level of its description? A view is now emerging according to which points in space and time can be though of as correlated quantum objects, just like electrons in a typical solid. Can spatial and temporal distances between these points then be described as the amount of entanglement between them? I believe that the next "big thing" will be the interaction between quantum information and gravity, hopefully resulting in a reasonable theory of quantum gravity.

In every epoch of scientific development, we have benefited from comparing the universe to the state-of-the-art technology of that epoch. In Newton's time the universe was understood to be a clockwork-like mechanism, possibly wound up by God at the beginning, but from then on running on according to the strict Newtonian laws of dynamics. In the next epoch, the universe was compared to a steam engine, producing heat as work is done, thereby evolving towards its final destiny, a thermal equilibrium in which no activity was any longer possible—the so-called heat death. We now live in the information age. Both physics and biology have contributed to a view of the universe as a giant computer,[1] and its evolution is seen to be nothing but processing of a certain type of information (genetic information in biology and quantum information in physics). It is clear that this picture has been very fruitful in

---

[1] It seems that a famous 11th century Persian mathematician and a poet, Omar Khayyam, had already beautifully anticipated this computing view of the universe in one of his poems from the collection known under the title "Rubaiyat":
But helpless Pieces of the Game He plays
Upon this Chequer-board of Nights and Days;
Hither and thither moves, and checks, and slays,
And one by one back in the Closet lays.

some research areas of science. However, whether it will help us reach the "holy grail of physics"—the ultimate understanding of the universe with one simple principle—is, of course, completely open to speculation[2]. Here your guess is definitely at least as good as mine.

---

[2]Here are recommend books by Deutsch and Tipler for some very speculative, albeit at the same time very exciting views on the future of physics.

Bibliography

Bell, J. (1987). *Speakable and Unspeakable in Quantum Mechanics*, Cambridge University Press, Cambridge, p. 2881. 2881.
Bhatia, R. (1997). *Matrix Analysis*, Springer, Berlin.
Brillouin, L. (1956). *Science and Information Theory*, Academic Press, New York.
Caves, C. M., and P. D. Drummond (1994). *Rev. Mod. Phys.* **66**, 481.
Cover, T. M., and J. A. Thomas (1991). *Elements of Information Theory*, Wiley Interscience, New York.
Csiszár, I., and J. Körner (1981). *Coding Theorems for Discrete Memoryless Systems*, Academic Press, New York.
Davies, E. B. (1976). *Quantum Theory of Open Systems*, Academic Press, London.
Deutsch, D. (1998). *The Fabric of Reality*, Penguin, London.
Everett, H., III (1973). The theory of the universal wavefunction, in *The Many-Worlds Interpretation of Quantum Mechanics*, edited by B. DeWitt and N. Graham, Princeton University Press, Princeton, NJ, Chapter 6.
Feynman, R. P. (1996). *Feynman Lectures on Computation*, edited by A. J. G. Hey and R. W. Allen, Addison-Wesley, Reading, MA.
Fuchs, C. A (1996). *Distinguishability and Accessible Information in Quantum Theory*, PhD thesis, University of New Mexico, Albuquerque, NM (lanl e-print server: quant-ph/9601020).
Garey, M., and D. Johnson (1979). *Computers and Intractability: a Guide to the Theory of NP-Completeness*, Freeman, San Francisco.
Gordon, J. P. (1964). *Noise at Optical Frequences; Information Theory, Quantum Electronics and Coherent Light*, Proc. Int. School Phys., Enrico Fermi, Course XXXI, ed. P. A. Miles, p.p. 156, Academic Press, New York.
Ingarden, R. S., A. Kossakowski, and M. Ohya (1997). *Information Dynamics and Open Systems — Classical and Quantum Approach*, Kluwer Academic, Dordrecht.
Kolmogorov, A. N. (1950). *Foundations of The Probability Theory*, Chelsea, New York.
Kraus, K. (1983). *States, Effects and Operations: Fundamental Notions of Quantum Theory*, Lecture Notes in Physics 180, Springer, Berlin, 1983.
Leff and Rex (2003). *Maxwell's Demon 2*
Lebedev, D. S., and L. B. Levitin (1963). Sov. Phys. Dokl. **8** 377.
Mackey, G. W. (1963). *Mathematical Foundations of Quantum Mechanics*, Benjamin, New York.

von Neumann, J. (1955). *Mathematische Grundlagen der Quantenmechanic*, Springer, Berlin, 1932; English Translation, Princeton University Press, Princeton, NJ.
Ohya, M., and D. Petz (1993). *Quantum Entropy and Its Use*, Texts and Monographs in Physics, Springer, Berlin.
Ozawa, M. (1984). *J. Math. Phys.*, **25**, 79.
Papadimitriou, C. H. (1995). *Computational complexity*, Addison-Wesley, New York.
Penrose, O. (1973). *Foundations of Statistical Mechanics*, Oxford University Press, Oxford.
Peres, A. (1993). *Quantum Theory: Concepts and Methods*, Kluwer Academic, Dordrecht.
Redhead, M. (1987). *Incompleteness, Nonlocality and Realism*, Clarendon Press, Oxford.
Reed, M., and B. Simon (1980). *Methods of Modern Mathematical Physics— Functional Analysis*, Academic Press, New York.
Schumacher, B. (1995). Phys. Rev. A **51**, 2738.
Shannon, C. E., and W. Weaver (1949). *The Mathematical Theory of Communication*, University of Illinois Press, Urbana, IL.
Stern, T. E. (1960). IEEE Trans. Info. Theory **6**, 435.
Tipler, F. J. (1994). *The Physics of Immortality*, Bantam Doubleday Dell, New York.
Tolman, R. C. (1938), *The Principles of Statistical Mechanics*, Oxford University Press, Oxford.
Vedral, V. (2002). Rev. Mod. Phys. **74**, 197.
Vedral, V. (2005). *Modern Foundations of Quantum Optics*, Imperial College Press, London.
Wehrl, A. (1978). Rev. Mod. Phys. **50**, 221.
Yamamoto, Y., and H. A. Haus (1986). Rev. Mod. Phys. **58**, 1000–1003.

Index

3-colorability, 132

additivity, 5
affine transformation, 5
alphabet, 62
atypical sequence, 7
atypical subspace, 58

basis, 18
BB84 protocol, 33
beam splitter, 25
Bekenstein bound, 157
Bell inequalities, 86
Bell states, 29, 44
Bell's inequalities, 83, 87
bipartite state, 38
bra (Dirac notation), 14, 30
braket, 30

Carathéodory, 123
Carnot cycle, 110
Cartesian coordinates, 14, 18
channel capacity, 72, 73
CNOT gate, 161
column vector, 16
complete complex vector space, 17
completely positive, 41, 95
completeness relation, 40
complex vector space, 17
conditional entropy, 8, 10, 59
conjugate, 15
conjugate transpose, 15
CP-map, 40, 41, 69, 95, 96
cryptography, 33, 140

data compression, 7, 55, 63
database search, 137, 147
dense coding, 45

density operator, 28
detector, 26
Deutsch's algorithm, 133
diagonal form, 22
Dirac notation, 14, 24
disentangled states, 85
distillation, 108, 114, 115
Donald's inequality, 61

eigenvalue, 21
eigenvector, 21
entanglement, 29
entanglement swapping, 48
entanglement witness, 93
entropy of observation, 58
EPR paradox, 81

factorization, 140
fidelity, 53, 106
formation of entanglement, 109, 114, 115
Fourier transform, 142
function of an operator, 24

generalized measurements, 40
God, 176
Grover's algorithm, 137, 147

Hadamard operation, 21
Hadamard state, 19, 21, 29, 33
Helstrom's discrimination, 54
Hermitian operator, 19, 23, 27
hidden variables, 82
Hilbert space, 17
Holevo bound, 71, 135, 154

identity, 20
inner product, 15, 16, 18, 24, 40, 53

instantaneous communication, 49
inverse, 20
inverse quantum Fourier transform, 144

Jamiolkowski isomorphism, 95
joint entropy, 8, 10

ket, 14, 30

Landauer, 31, 77, 78, 169
linear combination, 18
linear map, 95
linearly independent, 18
LOCC, 85, 110, 112, 127

Mach–Zehnder interferometer, 25, 102
map, 95
matrix, 16, 21
Maxwell's demon, 167
Maxwell–Boltzmann distribution, 6
mixed state, 28
mutual information, 8, 10, 12, 59, 68

no-cloning theorem, 31
norm, 15, 18
normal operator, 22
normalization, 15, 25, 35
NP-complete, 133

observable, 27
operation
    Hermitian, 27
    identity, 20
    inverse, 20
operator, 19
Oracle, 136, 147
orthogonal basis, 18, 20
orthogonal measurement, 27
orthogonal set, 18
orthogonal states, 15, 17
outer product, 19

partial trace, 35, 36
partial transposition, 97, 104
Pauli operator, 44, 161
Peres–Horodecki criterion, 97, 104
phase error, 161

phase estimation, 144
photon, 26
Pierls–Bogoliubov inequality, 72
positive map, 41, 95
postulates of information theory, 5
postulates of quantum mechanics, 27, 42
POVM (positive operator-valued measurements), 41
projection, 19, 27
projective measurement, 19
pure state, 28, 29
purification, 38

quantum Fourier transform, 142
qubit, 16, 20

relative entropy, 8, 9, 61–63, 65, 68, 69, 76, 115
relative entropy of entanglement, 115
row vector, 16

Sanov's theorem, 65
second quantization, 87
secure key exchange, 35
self-adjoint, 95
separable states, 85
Shannon, 4
Shannon entropy, 5, 7, 8
Shor's algorithm, 140
Socrates, 148
spanning set, 18
spectral decomposition, 22
spectral theorem, 22, 23
state, 27
subadditivity, 68
swap operator, 104

teleportation, 46
tensor product, 25
thermodynamics, 76, 110, 123
time complexity, 136
Toffoli gate, 161
trace, 35
transpose, 15, 41
triangular inequality, 68
type (of sequence), 62

type class, 62
typical sequence, 7
typical subspace, 58

unitary operator, 20, 23
unknown states, 31

vector space, 17

Werner states, 87